The BIOSYSTEMATICS of AMERICAN *CROWS*

by David W. Johnston

UNIVERSITY OF WASHINGTON PRESS · SEATTLE

1961

*This book is published with assistance from
a grant by the National Science Foundation.*

Copyright © 1961 by the University of Washington Press
Library of Congress Catalog Number 61-11576
Printed in the United States of America

ACKNOWLEDGMENTS

A comprehensive study of this magnitude and nature would obviously have been impossible without the cooperation of many individuals and institutions. Museums from which specimens were borrowed and/or visited and the individuals who made specimens available were: United States National Museum (Herbert Friedmann), Academy of Natural Sciences of Philadelphia (James Bond), American Museum of Natural History (Dean Amadon), Peabody Museum of Natural History (Philip S. Humphrey), Museum of Comparative Zoology (James Greenway and Ernst Mayr), Carnegie Museum (Kenneth C. Parkes), Chicago Natural History Museum (Emmet R. Blake), The Florida State Museum (Oliver L. Austin, Jr.), University of Georgia (Eugene P. Odum), Charleston Museum (E. Milby Burton), Museum of Zoology at Louisiana State University (George H. Lowery, Jr.), University of Michigan Museum of Zoology (Robert W. Storer), University of Kansas Museum of Natural History (Richard F. Johnston), Colorado Museum of Natural History (Alfred M. Bailey), Montana State College (Clifford V. Davis), University of Colorado (Gordon Alexander), University of Idaho (Earl Larrison), University of Washington (Frank Richardson), Charles R. Conner Museum at Washington State University (George E. Hudson), University of Oklahoma and private collection (George M. Sutton), University of Texas (Robert K. Selander), Los Angeles County Museum (Kenneth E. Stager), Dickey Collection, University of California at Los Angeles (Thomas R. Howell), Museum of Vertebrate Zoology (Alden H. Miller), College of Puget Sound (Gordon D. Alcorn), University of British Columbia (Ian McT. Cowan), Provincial Museum (C. J. Guiget), National Museum of Canada (W. Earl Godfrey), and Royal Ontario Museum of Zoology (L. L. Snyder).

Chandler S. Robbins kindly provided unpublished nesting data for Maryland and other states, and Allen J. Duvall made available migration records from the files of the United States Fish and Wildlife Service. Tape recordings of crow voices were obtained through the courtesies of Hubert Frings and P. P. Kellogg. Suggestions and assistance in field work in Washington were afforded by Gordon D. Alcorn and Garrett Eddy. K. C. Emerson kindly identified the Mallophaga collected from crows.

Special acknowledgments are due: to Frank Richardson, whose unselfish, helpful endeavors made field studies in Washington fruitful and pleasant ones, and whose editorial comments have been much appreciated; to Robert A. Norris and Eugene P. Odum, both of whom have been constant sources of encouragement, advice, and counsel throughout the entire project; and to Ernst Mayr for additional editorial suggestions.

These studies would have been impossible without the generous financial support afforded through three grants by the National Science Foundation. Field work in the summer of 1956 at the Savannah River Plant of the Atomic Energy Commission was abetted through a grant from the Southern Fellowships Fund. The administrative staffs of Mercer University and Wake Forest College were helpful in many ways toward the completion and publication of these investigations.

CONTENTS

1.	Introduction	3
2.	Methods	6
3.	Common Crow, *Corvus brachyrhynchos* Brehm	12
4.	Fish Crow, *Corvus ossifragus* Wilson	60
5.	Mexican Crow, *Corvus imparatus* Peters	71
6.	Cuban Crow, *Corvus nasicus* Temminck	81
7.	White-necked Crow, *Corvus leucognaphalus* Daudin	85
8.	Palm Crow, *Corvus palmarum* Württemberg	90
9.	Jamaican Crow, *Corvus jamaicensis* Gmelin	97
10.	Other Interspecific Relationships	100
11.	Summary and Conclusions	105
	Literature Cited	109
	Index	117

FIGURES

1. Approximate breeding distributions of three species of North American crows ... 11
2. Graphic expressions of measurements of adult male *Corvus brachyrhynchos* ... 38
3. Graphic expressions of measurements of adult female *Corvus brachyrhynchos* ... 42
4. Extreme and mean dates of breeding, *Corvus brachyrhynchos* and *Corvus ossifragus* ... 68
5. Graphic expressions of wing/tail ratios of selected crow populations ... 78
6. Approximate breeding distributions of four species of crows in the Caribbean Islands ... 83

TABLES

1. Measurements of *Corvus brachyrhynchos* ... 46
2. Measurements of adult male *Corvus ossifragus* from northern and southern portions of the range ... 69
3. Measurements of *Corvus ossifragus* ... 70
4. Measurements of *Corvus imparatus* ... 79
5. Wing/tail ratios in adult male crows ... 80

6.	Measurements of *Corvus nasicus*	84
7.	Measurements of *Corvus leucognaphalus*	88
8.	Measurements of *Corvus palmarum*	95
9.	Measurements of *Corvus jamaicensis*	99
10.	Weights of adult *Corvus brachyrhynchos* and *Corvus ossifragus*	101
11.	Occurrence of mallophaga on American crows	101

The BIOSYSTEMATICS of AMERICAN *CROWS*

1. INTRODUCTION

Among the more challenging instances of contemporary avian systematics and speciation are those concerning species which are closely related in morphological features, distribution, or ecology. The principal studies of such closely related species in the last decade have generally considered allopatric or sympatric forms which show distinctive morphological and sometimes ecological differences among one another. Studies of this nature have been carried out on titmice (Dixon, 1955), jays (Pitelka, 1951), meadowlarks (Lanyon, 1957), nuthatches (Norris, 1958), and others. Depending upon the group of species being studied, more or less attention was devoted to similarities and/or differences in color, measurements, and other definable characteristics. Taxonomists encounter more difficulty, however, in studies of obviously closely related species which do not always differ strikingly in their morphological features.

Such is the case in crows of the genus *Corvus*, for wherever these birds occur in the world, generally they are uniformly black, exhibit flocking behavior, have loud, raucous calls, stout bills and feet, and are renowned for their abilities in learning, mimicry, and intelligence. These similarities are, of course, too general, for indeed the various crow populations of the world have been relegated to distinct species, but the differences among these species are less striking than those which one usually encounters in most passerine types. In these crow species there are various degrees of sympatry and allopatry, small and large geographic ranges, and insular and continental types, so that studies of selected groups of crow species would help elucidate those factors which prevent interbreeding of sympatric or contiguous populations. In short, in this group of birds it is possible to select certain populations or groups of populations for special studies of speciation and reproductive isolating mechanisms.

The crows which inhabit the Western Hemisphere are particularly instructive in this regard, for on the continent of North America alone there are several similar-appearing species of crows, some of which are sympatric, others allopatric, some large, some small, some preferring one habitat, some another, and so on. Similarly, on various of the Caribbean Islands are other crow species, again exhibiting a wide range of ecological and morphological traits; some of these are overlapping, but others are species-specific. Aside from a consideration of those traits which can be utilized in delimiting a given species, one of the more interesting aspects of these American crows is the analysis of isolating mechanisms, because in the past studies of crows have emphasized the morphological features. In the newer approach to systematics, however, attention is focused upon the biological features which characterize each species and which prevent interbreeding between species.

This new systematics or biosystematics of American crows constitutes the bulk of the present investigations. Morphological characteristics are certainly not minimized here because variations in mensural features and the statistical analysis of these variations are basic to an understanding of any species. It is, however, the complex of biological characteristics which needs to be emphasized in these birds and used in conjunction with the morphological ones. As a result of the present study, one might say that the traditional species of American crows have now been defined on characteristics relating to their geographic range, habitat choice, voice, reproductive phenology, and others of an ecologic nature, in addition to the morphological characteristics obtainable from the usual museum specimen.

Currently, the consensus of American ornithological thought (A. O. U. Check-list, 1957; Miller, 1957) recognizes four species of crows as occurring in North America: Common Crow *(C. brachyrhynchos)*, Northwestern Crow *(C. caurinus)*, Fish Crow *(C. ossifragus)*, and Mexican Crow *(C. imparatus)*. In the Caribbean Islands the following four species are recognized (Bond, 1947): Cuban Crow *(C. nasicus)*, Palm Crow *(C. palmarum)*, White-necked Crow *(C. leucognaphalus)*, and Jamaican Crow *(C. jamaicensis)*. These eight species form a kind of unit or superspecies of all Western Hemisphere crows, a unit which presents interesting problems of species' differences, isolating mechanisms, and speciation. A more comprehensive treatment

Introduction

of the crows of the world in which intercontinental relationships could be explored more fully might be built upon this report in the future.

In this connection, mention must be made of the allegation by some that the Common Crow of North America is simply a subspecies of the European Carrion Crow *(C. corone)*. To my knowledge, no one has made a detailed comparison of these two forms, but Hellmayr (1934: 3) stated: "Comparison of a very large series of the European Crow in Field Museum shows the American Crow to be clearly conspecific." Other authors take sharp issue with this opinion by placing the Common Crow in the species *brachyrhynchos* (Meinertzhagen, 1926: 57, ff.), because there are at least significant voice differences (Laing, 1925: 35-36). A resolution of this problem would be forthcoming only after a careful analysis and comparison of characteristics of the two forms. Following the consensus of American authors, in this study the Common Crow is assigned to the species *brachyrhynchos*.

Van Tyne and Berger (1959: 370) in their discussion of "problem" species stated: "With the exception of a few isolated instances, so little is known about physiological, ethological, and ecological differences, especially among the "problem" groups, that data from these fields must be collected and evaluated critically if the usual taxonomic pitfalls are to to avoided." It is the purpose of the present analysis to evaluate accurately and correctly the ecologic relationships among the American crows which form one of these problem groups.

2. METHODS

Early in these investigations it became apparent that colors were less important in defining American crow populations than they are in some other corvids (see Pitelka, 1951: 199, ff.; cf. Vaurie, 1958), so elaborate color designations were not considered absolutely essential here. Even so, Ridgway (1904: 267, ff.) gave some color descriptions for the various American species, though somewhat greater emphasis was placed on measurements. Thus, for many years these two morphological features—color and size—provided the basis for the definition of American crows. No doubt these are basic species' characteristics, but it is also true, especially from a biological approach, that habitat, voice, and other features should be used in defining these crows.

The characteristics utilized in the present study for both species and subspecies are:

1. Mensural

a. Wing chord: this and the next measurement were recorded to the nearest half of a millimeter. Worn wings or tails were not measured.

b. Tail length: taken from the base of the middle rectrices to the tip of the longest rectrix.

c. Tarsus: the distance from the joint between tibiotarsus and tarsometatarsus to the distal edge of the most distal unbroken scale which crossed the bases of the three forward toes. This and the next measurement were recorded to the nearest tenth of a millimeter.

d. Bill: at first bill measurements were taken from the base of the bill at the forehead to its tip, but this measurement was quite unsatisfactory owing to the difficulty in locating a precise, uniform starting point. For this reason, all bill measurements were eventually taken, and are recorded in

Methods

the tables, from the anterior edge of the nostril to the tip of the bill. It is true, as Mayr, Linsley, and Usinger (1953: 129) stated, "the length of the bill, measured from nostril to tip, does not give the full length of the bill," but in these crows it was virtually impossible to locate a uniform point which marked the base of the bill.

2. Color

   a. Over-all: general color effect of the dorsum or ventrum, or sometimes a specific color of a restricted topographical position was recorded.

   b. Base of contour feathers: of especial use in the neck and nape of some species.

   c. "Scale-like Effect": a scalation or scale-like appearance of the upper back in certain species.

3. Habitat Choice. This feature, largely overlooked previously in defining these species, is of paramount importance in crows, and in some instances may be a rather effective isolating mechanism between two geographically sympatric forms.

4. Voice. Although some attention has been directed toward this feature in crows by previous authors (Amadon, 1944: 3), it was Mayr (1956: 113) who emphasized the need of modern techniques for analyzing crow voices and using such analyses for elucidating various knotty relationships. Especially did he have in mind the unclear relationship between *C. corone* and *C. brachyrhynchos*. More recently Davis (1958) has suggested various relationships among crows based upon voice characteristics; his interpretations will be discussed in detail later in this paper.

5. Geographic distribution and geographic isolation.

6. Weights. When available, the weights of crows from various populations are used to assist in the expression of size. This term "size" has been used so loosely among ornithologists that it needs to be standardized (see Connell, Odum, and Kale, 1960), and even though a large number of weights were not available for this study, probably weights would be better indicators of true size than wing or tail lengths. Even so, extreme caution must be exercised in using weights as statistical tools due to possible variations in fat.

7. Miscellaneous characters. Data for most American species are incomplete, but in some species one or more of the following features might be utilized in defining a given population:

   a. Shape of bill

   b. Reproductive phenology

   c. Behavior

d. Ectoparasites
e. Migration

Sex and age characters. No one has yet demonstrated unequivocal sexual dimorphic characters of crows either in the field or in specimens, in spite of many claims based upon size, behavior at a nest, voice, and the like (see Johnston, 1959, for a critique). Females tend to be smaller in weight and linear measurements, as will be indicated later, but there is much overlap even in these characteristics between the sexes.

It was Emlen (1936) who clearly demonstrated several morphological differences between adult and first-year Common Crows *(C. brachyrhynchos)*. The most apparent and easily recognized difference concerns wear and fading of the remiges and rectrices. Since the first-year birds retain these feathers, acquired by the bird while still in the nest, for at least fifteen or sixteen months, they undergo more wear and fading than comparable feathers in the adult. Hence, a first-year bird even in its first fall will exhibit this feature, whereas adults taken at the same time will have blacker, less worn tips of the remiges (especially the primaries) and the rectrices.

This feature alone is usually sufficient to distinguish adults from first-year specimens, but, as Emlen points out, the distal ends of the outer rectrices are also reliable age characteristics. These feathers on the first-year birds are generally more or less pointed, but a comparable adult feather would be more or less square-tipped.

Using principally these differences (wear, fading, and shape of feathers), it was possible to distinguish with virtually 100 per cent accuracy the two age groups of *C. brachyrhynchos*. It was also used accurately for *ossifragus, imparatus,* and *palmarum,* but some difficulty was encountered in determining the age of *nasicus, jamaicensis,* and *leucognaphalus*. In these latter species, if evidence was inadequate to determine the age correctly of a given specimen, that specimen was omitted from the study.

Even though hundreds of first-year specimens were available in museum collections, more than one half of these could not be measured accurately for wing and tail lengths due to the wear of these feathers. On these specimens, however, the tarsus and bill were measured because these "soft parts" apparently do not suffer the same degree of wear as the feathers. It must be made clear that none of these measurements from first-year birds is as standardized as those of an adult, simply because the first-year birds are still growing to the adult size. A careful, sea-

Methods

sonal study of these immatures, involving measurements alone, would probably show that summer specimens just out of the nest are smaller than spring specimens about one year old. Granted that these variations might be true, the present study did not concern refinements of this magnitude, so that all first-year specimens of a given species were placed in the same sample though segregated according to sex.

Statistical formulae. The following formulae were used uniformly for calculating statistical expressions in this paper. For samples of ten or less only the extremes and means were ascertained.

1. Standard Deviation

$$S.D. = \sqrt{\frac{\sum x^2}{N} - M^2}$$

(Since this statistic was calculated for sample sizes varying from 10 to 89, N was used uniformly instead of N-1.)

2. Standard Error of the Mean

$$S.E._m = \frac{S.D.}{\sqrt{N}}$$

3. Coefficient of Variability

$$C.V. = \frac{S.D. \times 100}{M}$$

4. Pooled standard deviation

$$S.D._p = \{[(N_1)(S.D._1^2) + (N_2)(S.D._2^2) + (N_1)(M_1^2) + (N_2)(M_2^2) - (N_1 + N_2)(M_p^2)]/(N_1 + N_2)\}^{\frac{1}{2}}$$

(MacArthur and Norris, in press)

5. Standard Error of the Pooled Means

$$S.E._{mp} = \frac{S.D._p}{\sqrt{N_1 + N_2}}$$

Selection of specimens. Only specimens of *brachyrhynchos* and *ossifragus* known to be from the immediate breeding locality were utilized here. In general, this meant specimens taken between April and September, but for extreme northern or southern localities these limits were extended so that an accurate breeding sample could be obtained. For example, the Common Crow in Florida begins to breed in late January (see Figure 4), so the breeding sample for this species in Florida began with January. For some of the northern states, a September specimen might represent an early migrant from farther north. Every precaution was taken, then, to include in this investigation only birds believed to be from the breeding area, and all dubious specimens were discarded from the analyses. Even so, one cannot be absolutely certain that some of these samples did not include one or more specimens which failed to migrate in the spring or summer to its natal ground. Such specimens certainly composed an insignificant portion of the 2269 specimens examined in this study.

Winter specimens of *brachyrhynchos* and *ossifragus* are not included at all, primarily because, in the absence of a banded bird, one cannot be absolutely certain of a specimen's breeding locality. Undue importance has been assigned, I believe, to the almost compulsive fetish of assigning a trinomial to a winter crow specimen. We do know from banded birds that many crows in Oklahoma and Texas in winter actually bred or were raised in Saskatchewan or Manitoba, and that some wintering crows in the southeastern United States were raised in eastern Canada or the northeastern states. But it does not follow that a large winter specimen, from North Carolina for example, must be a representative from some far northern population. As will be clearly demonstrated later, such a specimen could easily represent a large individual of the local breeding population.

In order to identify unbanded winter crow specimens, one should consult the several tables presented in this investigation, and on the basis of these measurements attempt to locate as accurately as possible the breeding population which most closely fits.

Little is known concerning migration of Mexican crows, but there seems to be no evidence that they do migrate. Even so, specimens collected between October and March were not used for measurements, just to be "on the safe side." No doubt, the Caribbean crows do not migrate, so a specimen taken at any time of the year could hardly be very far from its breeding site. For this reason, all specimens of the Caribbean forms were utilized in these investigations.

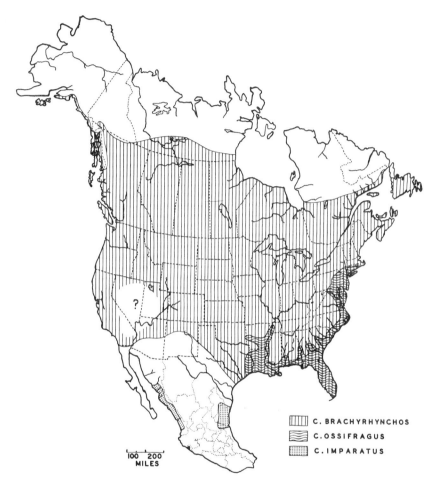

Figure 1. Approximate breeding distributions of three species of North American crows.

3. COMMON CROW

Corvus brachyrhynchos Brehm

Corvus brachyrhynchos Brehm.

Geographic Range
United States: specimens examined, 1204: from Alaska, 86; Washington, 176; Oregon, 43; California, 107; Idaho, 29; Montana, 30; Wyoming, 6; Arizona, 6; Colorado, 10; New Mexico, 12; North Dakota, 40; South Dakota, 1; Kansas, 10; Oklahoma, 10; Texas, 14; Minnesota, 4; Wisconsin, 12; Iowa, 4; Arkansas, 9; Louisiana, 12; Illinois, 13; Michigan, 35; Ohio, 7; Kentucky, 12; Tennessee, 13; Mississippi, 6; Alabama, 8; Florida, 131; Georgia, 145; South Carolina, 26; North Carolina, 11; Virginia, 21; West Virginia, 8; Pennsylvania, 12; Maryland, 8; District of Columbia, 6; New Jersey, 14; New York, 52; Connecticut, 15; Rhode Island, 2; Massachusetts, 27; Vermont, 1; New Hampshire, 4; and Maine, 6; others reported in the literature from Nevada, Utah, Missouri, Indiana, Nebraska, and Delaware.
Canada: Specimens examined, 494: from Nova Scotia, 8; New Brunswick, 3; Newfoundland, 6; Prince Edward Island, 2; Quebec, 31; Ontario, 83; Manitoba, 29; Saskatchewan, 18; Alberta, 41; British Columbia, 271; Yukon, 1; Keewatin, 1.
Mexico: specimens examined, 2: from Baja California (nonbreeding).
Bermuda: specimens examined, 1 (breeding?).

Specifically, the range of the Common Crow during the breeding season (Figure 1) extends northward (east of the Coastal

Range in British Columbia) to Dezadeash Lake, Yukon (although Bishop [1900: 81] reported crows observed on the Porcupine River, this extremely northern record seems dubious); Simpson, Grandin River, Rae (Preble, 1908: 406), and Artillery Lake (Harper, 1953: 77), MacKenzie; Nueltin Lake, Windy River (Harper, *ibid.*), and Eskimo Point (Sentry Island), Keewatin; Churchill and York Factory, Manitoba; Fort Severn, Attawapiskat River, and Moose Factory, Ontario; Lake Mistassini, Leroy Lake, Mile 72 (Q. N. S. and L. Railway) (Harper, 1958: 102), Mingan Islands, Natashquan, and St. Paul River (Harper, *ibid.*), Quebec; and throughout Newfoundland, except the northern peninsula (Peters and Burleigh, 1951: 309). Along the northwestern coast, *Corvus brachyrhynchos caurinus* extends northward in Alaska to at least Valdez and westward to Kodiak Island and Kukak Bay (Gabrielson and Lincoln, 1959: 620).

The southern breeding limits extend to northern Baja California (egg set in American Museum Natural History, Vallecitos, 20 miles N. Ensenada, Mexico, April 7, 1920); central Arizona (Gila County; a specimen (United States National Museum 79686) labeled "Apache, Arizona" which would be in Cochise County probably refers to "Fort Apache" in the central part of the state); north central New Mexico (Santa Fe and San Miguel counties; Bailey, 1928: 494); northern, central, and south central Texas (Wolfe, 1956: 51-52); and the Gulf of Mexico.

(These detailed locality records are taken from specimens unless otherwise denoted by citations from the literature. In some instances they represent notable extensions of the known breeding range of this species as recorded in the A. O. U. Checklist (1957).)

Habitat. So much has been written about the habitats where Common Crows occur that it is difficult to make an accurate, general, and concise statement with reference to habitat choice. Virtually all the state bird books treat this subject at least briefly, and one phrase seems to stand out as best describing the habitat preference—farmland with adjoining woodland. It is this combination of ecologic situations in the environment which best suits the needs of the Common Crow, the farmland providing food and the woodland or forest offering cover and nesting sites. There are, of course, exceptions to this statement—instances of Common Crows being found on the prairies where there are no trees and of their occasional occurrence in fairly heavily forested regions where there is virtually no agricultural land—but there is considerable evidence to support the contention that the Common Crow reaches its peak of abundance in agricultural lands where there are adjacent woods.

Recently Stewart and Robbins (1958: 221) showed that the Common Crow in Maryland reached its greatest breeding density (0.6 pairs/100 acres) in mixed forests and brush habitats where there were adjacent small agricultural areas and abandoned fields. In various agricultural habitats where there was little cover due to widespread clean-farming practices, the breeding density was considerably less (0.1 pair/100 acres).

Good (1952: 2) in his life history study of this species reported: "The crow is more or less closely associated with man's activities." This author showed on several maps that the center of abundance for the Common Crow in the United States was in the great wheat and corn areas of the central states. Emlen (1940), although writing about midwinter distribution in California, reported that even in forested areas, centers of greatest abundance were along the river valleys where there was some agricultural land. Similar patterns of local distribution may be found in other parts of the range. For example, in the Okefinokee Swamp of Georgia where there is virtually no cultivated farmland, Common Crows are quite rare (Hebard, 1941: 61) but Fish Crows are not. And, in the heavily forested mountains of north Georgia, Common Crows are almost entirely limited to the mountain valleys where there are some agricultural lands.

Historically, several authors have indicated that the Common Crow, although present before the advent of white men in North America, has increased in numbers where agricultural practices have increased. Taverner (1919: 156) stated that "whereas the raven retreats before the advance of civilization, the Crow increases. . . . Without doubt the Crow has increased enormously in the country since the removal of the forests, and probably its advent in eastern Canada was coincident with the arrival of the white man." For western Canada he reported (Taverner, 1926: 259): "It is an open-country bird and . . . in the west where prairies and open spaces were the original condition of the country, it probably has always occurred, but it has increased enormously with civilization." Bent (1946: 226) wrote: "Even before white man came to America it was well known to the Indians and every tribe had its name for this bird, which was such a conspicuous creature of their environment." Finally, Good (1952: 22) related: "That the crow was not common in early times over those portions of its range where it is now most abundant is attested by many early writings. It was scarce in Wisconsin in 1855 but became abundant by 1890. The crow is not a forest bird like the raven, nor is it a prairie species. It thrives best where there are trees for nesting and roosting and open fields for feeding."

Voice. To describe the usual call of the Common Crow as a loud *caw* is an extreme oversimplification, but even this brief description provides a basis for comparison with other American species (cf. Fish Crow). In order to expand this description, it is difficult to know where to start, for virtually all ornithological writings which deal to any extent with the Common Crow have something to say, at least briefly, about its vocalization. Bent (1946) made an attempt to summarize some of the various calls of this species, devoting two pages to this subject, and Mathews (1921: 47-48) described in musical terms the usual *caw*. No less than ten different kinds of notes or calls are described by Good (1952: 36 ff.), many of these being slight variations of the *caw*. The most common of these was a *ca-a-a-a*, descending in pitch and varying with the rapidity of delivery. This latter qualification should be emphasized because even the most casual field observer can detect pitch variations depending upon the direction of the calling bird's head with respect to the observer and the speed at which the calls are given. In addition to these variations, there are possible differences attributable to sex and/or age, but, as yet, such variations have not been clarified.

With the perfection of highly technical recording equipment for field use, we have seen in recent years the development of this tool for ornithological research, and at least two significant studies have dealt with voices of crows. One of these, by Davis (1958) was a compilation of audiospectrograms, the analysis of which led to a comparison of different crow populations on the basis of voice. For the Common Crow his recording was of the simple "caah," and by using this recording, Davis demonstrated some interesting differences among this species, the Fish Crow, and the Mexican crows. Using a different approach, Frings, *et al* (1958) recorded voices of the Common Crow with the idea of (1) using these as a repellant and (2) comparing responses with some of the European crows. He was careful to state (p. 126) that "no attempt was made to test all the calls of the Eastern [Common] Crow." Four calls, analyzed from field recordings, were described in some detail: (1) the assembly call, (2) the alarm call, (3) ordinary cawing, and (4) begging cries of young nestlings. An even later report by the Frings (1959) has suggested some explanations for the responses obtained from certain calls of this and other crow species. All of these investigations are certainly instructive and contribute toward an understanding of Common Crow calls, but one of the effects of these studies has been the stark realization that this entire field of vocalization needs intensive study.

Morphology. In color, the Common Crow is rather distinctive when it comes to a comparison with the other North American and Caribbean species. The occiput and crown is lightly flecked with violet blue on a background of black, this area contrasting fairly sharply with the nape, which is a uniform dull black. This flecking on the head is more apparent on some specimens than on others. The upper back, scapular, central and lower back feathers are metallic violet and show the scale-like appearance alluded to by several authors (see discussion below). In some lights, there is a gloss of reddish violet to the primary coverts, scapulars, and secondaries, or there might be a slight greenish cast to the primaries. Ventrally, by way of contrast, the throat and neck are generally black with a wash of metallic violet. The chest and flanks are also washed with violet on a black background, and usually give a mottled appearance. The central abdomen and belly feathers are dull black without any violet luster.

Of all the species of crows studied here, specimens of *brachyrhynchos* most closely resemble *nasicus* in color. Certain distinctions can be made, however, because *nasicus* has less of the scale-like appearance to the back. There is, furthermore, less contrast between the head, nape, and back, because in *nasicus* the nape is about the same shade of deep violet as the back of the head. The underparts of these species are not easily distinguishable, but *nasicus* is usually more glossy below than is *brachyrhynchos*. (There are, of course, significant size differences which one could use in distinguishing specimens: *brachyrhynchos* has larger wings, tail, and tarsus, whereas *nasicus* has a longer bill.)

This scale-like appearance of the dorsum deserves further explanation here since its occurrence can be used to distinguish several species. Dorst (1947: 72), in his key to the species of *Corvus*, described this condition as "squamous; edge of the feathers clearly defined" as versus "feathers with non-coherent barbs, their edges not very defined." Though this contrast would seem to be clear-cut, the term "non-coherent" barb is a relative one.

The reason for this over-all scale-like appearance becomes evident upon microscopic examination of the individual contour feathers involved. A 50-mm. feather from the scale-like back of a Common Crow shows distally on the barbs a 1-2 mm. band where the barbules fail to interlock, whereas, comparatively, a similar feather from a Fish Crow, lacking the conspicuous scale-like appearance, has a band 5-7 mm. wide where the barbules fail to interlock. Individually, the barbules grow shorter toward the distal ends of these feathers and barbicels are either absent or greatly reduced in number distally in this band where the barbules do not interlock. These two features, especially the shorter barbules, are responsible for the "non-coherent barbs" of Dorst or, rather, the frayed-out appearance of the end of the feather.

A rather sharp contrast is presented, then, between the barbules which do and those which do not interlock; at least in the Common Crow this is true. And apparently the shorter the distance of these noninterlocking barbules, the sharper the contrast and the greater *(in toto)* the scale-like appearance.

Subspecies:

It has been customary for most American taxonomists to recognize at least four geographic races or subspecies of *Corvus brachyrhynchos,* and at least two others have been proposed at one time or another. The most recent consensus, viz., the A. O. U. Check-list of North American Birds (1957), listed the following:

1. *C. b. brachyrhynchos.* An eastern population extending westward in North America to northeastern Alberta in Canada and to about 100° West longitude in the United States and southward to approximately the Ohio River. Here it presumably intergrades with *C. b. paulus.*

2. *C. b. paulus.* A more southerly form occupying the region from about the Ohio River southward to the Gulf of Mexico and the northern border of Florida and westward to eastern Texas and Arkansas.

3. *C. b. pascuus.* Peninsula of Florida.

4. *C. b. hesperis.* A western population occupying southwestern Canada (British Columbia, Central Alberta and Saskatchewan) and the western United States, southward to approximately the Mexican border. It presumably intergrades with *brachyrhynchos* and *paulus* along its eastern border.

In addition to these currently accepted forms, Phillips (1942:

573-75) proposed a subspecies from the southwestern United
States, *C. b. hargravei*. Also, the Northwestern Crow, *C. b.
caurinus*, was at one time (A. O. U. Check-list, 1931) classified
as a subspecies of the Common Crow, although the 1957 Checklist recognized it as a distinct species. None of these six forms
can be distinguished on the basis of color alone, in spite of
Davis' contention (1958: 151) that *caurinus* differs in dorsal coloration from *hesperis*. The differences among them are relative
differences in size, a situation which has resulted in the expression "millimeter races."

Ridgway (1904: 269, ff.) stated that *C. b. pascuus* differed
from *C. a. americanus* (= *C. b. brachyrhynchos)* by being smaller, except for the bill and feet. In turn, *hesperis* differed from
americanus by being decidedly smaller, with bill relatively
smaller and more slender. And Howell (1913: 199-200) described
the subspecies *paulus* on the basis that it was "decidedly smaller
than *Corvus b. brachyrhynchos*, with a much slenderer bill.
Nearest to *Corvus b. hesperis* but with shorter wing and slightly
larger bill."

Before one can effectively tackle the problem of subspeciation
in this species, a few purely mechanical manipulations of the
specimens must be effected; unfortunately, as will be emphasized
below, these procedures have not always been followed by taxonomists in their classification of crow populations. Of paramount
importance is the proper sorting of specimens according to *both*
sex and age. If this is not done, one might, for example, erroneously compare adult and first-year birds; the conclusions
from such a comparison would obviously be invalid. Another error has been the failure to include in an analysis only breeding
birds or birds taken from the known breeding locality, and both
Howell (1913) and Phillips (1942) used specimens which were not
known to represent the breeding population from the locality
under consideration.

Unfortunately, Howell's description of *C. b. paulus* is not
very exemplary in this respect due to the heterogeneity of his
sample. While working through the crow specimens in the U. S.
National Museum, I had the opportunity of re-examining the
material upon which Howell's study was based, and found that
his specimens were not segregated properly according to age.
It is true that morphological differences between adults and
first-year birds were not known at the time when Howell described *paulus*, but this does not rectify the errors extant in his
data. Two of the presumed four adult males which he measured

from Autaugaville, Alabama, were, in fact, first-year birds. Similarly, his material from Texas, South Carolina, District of Columbia, Mississippi, and Louisiana was entirely of first-year birds, the measurements from which he grouped with that of adult birds from Virginia and Alabama.

Another rather serious flaw in his data was the inclusion of birds not definitely known to be from the breeding groups. In his total number of twenty-four specimens examined were several birds taken in November and February, and certainly these specimens could have been migrants from another breeding population. To compound confusion, some of the birds taken in February were also first-year birds.

After specimens have been properly grouped according to sex, age, and breeding locality, one can then proceed to analyze their measurements. Since there are four sex–age groups and since four standard measurements have been made, it is apparent that there are sixteen possible mensural differences between two populations, for example, bill of adult males, bill of first-year males, wing of adult females, and so on. In practice, however, it is generally unwise to consider the first-year birds in taxonomic considerations because, with the exception of only a few specimens, the wing and tail feathers were so worn and faded that measurements of them would likely be quite inaccurate. Also, first-year birds, which are generally smaller than adults, would not be accurate biological representatives of a feral population for the systematist. For this reason, in the following analyses only adult birds are considered, but measurements of first-year birds, when available, are given for reference in Table 1.

In recent years when avian taxonomists have devoted their energies toward the designation of subspecific populations based on mensural characters, rather significant questions have been raised as to the application and interpretation of statistics. Especially is this true when it comes to deciding the degree of difference permissible between populations which are supposed to be subspecies. One such application is the "seventy-five per cent rule" which enables the taxonomist to recognize a dividing line between populations whose measurements overlap. Various interpretations of this rule have come into being (see Mayr, Linsley, and Usinger, 1953), and the most commonly used expression of the rule was discussed by Amadon (1949) who concluded (p. 258) that one "expression of the rule, 97 per cent [different] from 97 per cent, is more accurate . . . when comparisons are based directly on specimens (or their measurements)." Actually this interpretation is based upon the graphic method of statistical analysis as described by Dice and Leraas (1936). Of paramount importance in this method is overlap (or lack of overlap) between bars representing twice the standard errors (hence, about 95 per cent) of the means of the various populations. If these standard errors overlap for a characteristic under consideration, one would consider no significant difference between the populations, and vice versa. This comparison (95 per cent from 95 per cent) is somewhat less stringent than Amadon's interpretation, and is used here principally because the standard errors may be computed and expressed with ease.

When considering a species which has such a large breeding range as does the Common Crow, an investigator should utilize samples from all parts of the range when available, but, in this species, so many specimens were available that it became necessary to arrange specimens in groups. Such a grouping was in part arbitrary, but was ultimately based upon two factors: (1) the number of specimens available, and (2) an attempt to use natural geographic or ecologic units such as the middle Atlantic states or the northern "prairie" states. Thus, the large number of specimens from such states as California or Florida permitted me to consider these states as distinct units, but so few specimens were available from Arizona, New Mexico, Colorado, and Oklahoma that the specimens from these states were combined into one unit. If this procedure had not been followed, it would have been impossible to have large enough sample sizes for reliable statistical computations and comparisons.

So far as the natural geographic or ecologic units were concerned, to take one example, birds from Montana and Wyoming were considered to be closer in their ecologic relationships to birds from the Dakotas than to birds from northern Idaho and eastern Washington. Admittedly, some of the units utilized here do not necessarily demand close ecologic relationships of the birds in each unit, but their inclusion into these units again facilitated statistical manipulations. Since so many units were used, it is felt that any "errors" incurred by placing one state's birds in the "wrong" group would be negligible.

Based upon sample size and geography, then, the following units or groups were utilized:*

---

*The precise number of specimens from each unit, grouped according to measurement, sex, and age may be found in Table 1.

1. Quebec, New Brunswick, Nova Scotia, Prince Edward Island, Newfoundland
2. Maine, Rhode Island, Connecticut, Massachusetts, New York, New Jersey, Pennsylvania
3. Maryland, District of Columbia, Virginia, West Virginia, Tennessee, Kentucky, North Carolina, South Carolina
4. Georgia
5. Florida
6. Ontario and Manitoba
7. Michigan, Ohio, Illinois, Wisconsin, Iowa, Minnesota
8. Texas, Louisiana, Alabama, Arkansas, Mississippi
9. Saskatchewan, Alberta, Yukon, Keewatin
10. Montana, Wyoming, North Dakota, South Dakota
11. Colorado, Oklahoma, New Mexico, Arizona
12. California
13. Oregon
14. Western Washington (King County and southward)
15. Western Washington (north of King County)
16. Interior Washington (east of the Cascades), interior British Columbia, and Idaho
17. Coastal British Columbia
18. Alaska

The states included in groups 3 and 8 (and possibly 4) above coincide more or less with those used by Howell in describing the subspecies *paulus,* and group 11 included some birds described as *hargravei* by Phillips.

Measurements of these groups are expressed graphically in Figures 2 and 3. Of course, it is rather arbitrary to arrange these groups from north to south; they could have been just as easily arranged from east to west. Regardless of their arrangement, the critical feature is to examine them by comparing contiguous populations and ultimately trying to decide whether population A and population B, or groups of populations, are sufficiently different from one another to warrant their recognition as subspecies.

Another feature of these crow populations needs to be borne in mind when the matter of subspeciation arises. Although there are no color distinctions, even with measurements we have more to consider here than is the usual case among birds. A cursory glance at the literature on avian systematics will reveal many subspecies described on a single difference—wing length, tail, or bill, these differences frequently being based on mean lengths from small samples. These might be quite valid characteristics for a given species, but the present analysis on crows entails a more intensive and extensive study involving not only hundreds of specimens but also four important mensural characteristics of each sex and age group. Within the framework of this fact, it is desirable to reflect on a statement by Van Tyne and Berger (1959: 355-56) "The *wise* use of trinomials, therefore, may be advantageous for indicating the relationship among different populations. A conservative treatment of trinomials seems indicated: use those that serve a purpose; discard those that are poorly conceived." Attention should also be drawn to a statement by Amadon (1950: 497):

> We can then only study the amount of variation in the group in question and decide arbitrarily whether a particular isolated form is best considered a species or a race. . . . Although as a rule relationships are perhaps best expressed by listing geographical representatives as subspecies whenever possible, great caution must be used in difficult genera containing many similar species. On such grounds one may seriously question Hellmayr's decision that the Northwestern Crow, *C. caurinus,* is a race of the Fish Crow, *C. ossifragus,* or that the American Crow, *C. brachyrhynchos,* is a race of the European Carrion Crow, *C. corone.*

It would seem to me that a valid subspecies population of crows should be one that differs from another subspecies in more than one of its characteristics, and preferably about one-half of the eight possible differences among the adults. Then, too, the distribution of these differences between the sexes is of importance because it seems poor judgment to consider a subspecies as valid when only one of the sexes can be identified or where only a scant percentage of the characteristics are significantly different. Some taxonomists refer to these latter cases facetiously as "weak races," but if a subspecies is scientifically valid and serves a useful purpose in the biology of a species, there should be no need for such expressions as "weak race" or "good subspecies."

C. brachyrhynchos pascuus. Of all the populations of *C. brachyrhynchos* examined, the birds from the Florida peninsula are the most distinctive in size (Figures 2 and 3). If one compares these birds with specimens from adjacent areas (Georgia, Alabama, Mississippi, etc., or really, the entire proposed range of *paulus)*, it is apparent that the crows from Florida are significantly larger in tarsal and bill lengths, and tend to have longer wings and tails but not always significantly so (adult males' wings and adult females' tails). In at least three fourths of this form's characterictics it is possible to distinguish it from adjacent populations, and in tarsus and bill *pascuus* is even significantly larger than all other populations of *C. brachyrhynchos*.

From these data one finds a strong argument for the retention of this Florida population, *pascuus*, as a valid subspecies. To state that the range of this form is limited to Florida must not be taken too literally because there is no *sharp* boundary between *pascuus* and more northerly populations. Specimens from southern Georgia, though not as large as the Florida specimens, generally fall into clines leading toward the extreme and distinctive size of the Florida population, but are closer to more northerly populations. Nonetheless, one must not lose sight of the fact that adult specimens from Florida are significantly different from specimens from Georgia in seven out of the eight standard mensural characteristics.

By comparing seven Common Crows from Pennsylvania, Virginia, and Georgia with five specimens from Florida, Bailey (1923) concluded that the Florida subspecies was not "worthy of a place in the the new check list" (p. 149). He claimed that "as a whole, both forms average up about the same" in measurements, and that other proposed differences were negligible or inaccurate. It is obvious that the more extensive measurements discussed in the present paper clearly negate Bailey's conclusion.

C. brachyrhynchos paulus. Howell (1913) characterized this subspecies by stating that it had a "much slenderer bill" than *C. b. brachyrhynchos*. Evidently "slender" bill referred to depth of bill at the nostril, for a comparison between Ridgway's data (1904: 267) and that of Howell would be as follows:

		Depth of bill at nostril	Exposed culmen
Adult male	Ridgway	17.5-20.5 (19.5)*	48-53.5 (51.5)
	Howell†	17 -18 (17.5)	47-54 (51)
Adult female	Ridgway	17.5-19.5 (18)	45.5-50 (48)
	Howell†	16 -16.5 (16.2)	45.5-46 (45.8)

*Means are given in parentheses.
†See earlier comments on the heterogeneity of Howell's sample.

Although these data might have been impressive to Howell, they are not precisely comparable because Ridgway included measurements from birds throughout the entire eastern United States, north of Florida, whereas Howell examined a small sample from Alabama alone. Furthermore, it is evident that both of these investigators used rather small samples (seven birds for Howell). The serious error in the inclusion of first-year birds by Howell has already been mentioned, and it was probably the smaller size of these birds which contributed to Howell's belief that *paulus* was a significantly smaller geographic race.

Howell further proposed that the distribution of *paulus* should be from the District of Columbia and southern Illinois south to South Carolina (probably Georgia), Alabama, Mississippi, Louisiana, and southeastern Texas. So that the more abundant data in the present study might be compared more effectively with Howell's data, measurements from groups 3, 4, and 8 above were combined by using the pooling formula of MacArthur and Norris. This combination of groups would then be the representative sample from the geographic area mentioned above as the proposed distribution of *paulus*. Thus, in Figures 2 and 3, with the exception of measurements of adult females' wings and tails (too few specimens from the "Tex.-Miss." sample), the data for birds in the proposed range of *paulus* are titled, "Md.-Ga.-Tex.", being, therefore, inclusive of most of the southeastern United States.

When one examines these data, as given graphically in Figures 2 and 3, it can be seen that specimens of *paulus* differ significantly from specimens farther north (the "Mich.-Iowa," and "Me.-Pa." samples) only in the adult males' wings. Wings of adult females are different from the "Me.-Pa." sample but not from the "Mich.-Iowa" sample. As far as tail, tarsal, and bill lengths of either sex are concerned, it is not possible to distinguish statistically *paulus* from the more northerly populations. Indeed, there are no consistent clines in these measurements, for adults of *paulus* tend to have larger tarsi and smaller bills when compared with northeastern crows, but in comparison with specimens from "Mich.-Iowa," the bill is larger.

No doubt many of the "large" winter specimens of crows from the southeastern states identified as *C. b. brachyrhynchos* by other taxonomists, and thus presumably from areas farther north, were in fact simply large examples of the local breeding population. This does not mean that the more northerly populations do not migrate southward in winter, but it does mean that the only certain way of allocating most winter specimens to a more northerly population is through banded birds.

In summary, the wing of adult males of *paulus* is the only characteristic (one out of a possible 8 in both sexes) which could be used unequivocally to identify this subspecies, all other characteristics showing more or less overlaps with adjacent, more northerly populations. On such exceedingly tenuous grounds, it seems unwise to recognize this intermediate population as a distinct subspecies.

C. brachyrhynchos hargravei. Since so few breeding specimens were available from Arizona and New Mexico, it was thought advisable to include with these some specimens from Colorado and Oklahoma, even though these last two states were not included in Phillips' original description of the range of *hargravei*. It would seem to be wiser to have this larger sample from a somewhat larger geographic area rather than to draw conclusions from an exceedingly small sample.

For wings and tails, in a general way, the data in Figures 2 and 3 lend support to Phillips' contention that specimens from this southwestern region tend to have long wings and tails like some specimens taken from eastern and more northerly crow populations. Even so, the "Colo.-Ariz." samples are not significantly different from geographically adjacent populations to the north and east. The same general conclusion may be made with respect to tarsal and bill lengths. In both sexes, the bills and tarsi may be longer or shorter, depending upon the population (northern or eastern) with which the comparison is made, but in any event there are no consistent clines in measurements for the two sexes.

Since the crows of this region are not significantly different from any adjacent populations according to the present conclusions, this southwestern population hardly warrants recognition as a distinct subspecies.

C. b. brachyrhynchos and *C. b. hesperis*. When one examines the data in Figures 2 and 3 for the eastern and western populations, it is immediately apparent that there are many variations in trends or clines, very few of which are consistent. For example, in adult males, the wings may be longer in southern populations ("Colo.-Ariz.") or shorter (the southeastern United States); on the other hand, bills and tarsi tend, over-all, to be larger in the more southerly populations, thus illustrating Allen's rule. The most clearcut clines are to be seen in the Canadian populations when traced from east to west, because birds from "Sask.-Alta." tend to have the shortest wings, tails, tarsi, and bills when compared with the "Que.-Newf." samples. When, however, these mensural data are critically analyzed for "breaks" or significantly different measurements, confusion again reigns, for such "breaks" when they do occur might be found between "Sask.-Alta." and "Ont.-Man." or between "Ont.-Man." and "Que.-Newf." depending upon the specific measurement under consideration. Under these conditions of overlapping measurements most specimens could not be correctly associated with a specific geographic location, so biologically it would be best to consider all these crows as belonging to the same subspecific population, along with those from the eastern and midwestern United States. Thus, the proposed range of *C. b. brachyrhynchos* would extend from the eastern coast of North America westward to Alberta, Montana, Wyoming, and Arizona, and southward to the Gulf of Mexico.

Migration data lend support to this new proposed range of *C. b. brachyrhynchos*. The breeding populations of crows from Alberta, Saskatchewan, and Manitoba winter from Montana and North Dakota southward to Colorado, Oklahoma, and Texas. Similarly, crows from the more easterly Canadian provinces are known to winter in the midwestern and eastern United States. This feature, plus the absence of distinctly different populations (based on measurements), leads to the conclusion that all the Common Crows east of a line drawn from Alberta approximately southward to Arizona (with the exception of Florida) should be regarded as one subspecies.

Common Crows breed sparingly in the Great Basin region, so until sufficient numbers of specimens are available from southern Idaho, Utah, and Nevada, their subspecific allocation must remain undetermined.

Specimens from California, Oregon, interior Washington (east of the Cascades), northern Idaho, and interior British Columbia (east of the Coastal Range) are noticeably different from those taken in localities farther east, and here I am proposing that the range of the western subspecies, *hesperis*, be restricted to these geographical areas. Not only are specimens from California significantly different in many measurements from those taken in Alberta (5 out of 8 instances), but also there is a strong tendency for the interior British Columbia birds to be different from those taken in Alberta.

The *"Corvus caurinus"* problem

Ever since Baird described the Northwestern Crow as a distinct species in 1858, there has been much confusion in delimiting its range and characteristics, and numerous differences of opinion have arisen regarding its validity as a species. So confused was the picture that Baird, Brewer, and Ridgway (1875: 250) believed that "*C. caurinus*, according to Dr. Cooper, breeds down to southern California . . ."; of course, it is now obvious that these investigators were really referring to the western form of *C. brachyrhynchos*, for they stated that it preferred "inland districts to shores and bays" and gave several nesting localities (San Diego, Visalia, and Santa Cruz, California). In the early part of the twentieth century, however, ideas began to crystallize as to the range of *caurinus*, and it was the consensus then that this Northwestern Crow ranged from Alaska down the coast of British Columbia southward to Puget Sound in Washington. Later workers (Jewett and Gabrielson, 1929: 30; Gabrielson and Jewett, 1940: 427; Kitchen, 1949: 178) expressed the belief that *caurinus* occurred along the Olympic Peninsula of Washington to the mouth of the Columbia River. In fact, Jewett and Gabrielson were of the opinion that *C. caurinus* was *the* crow of the Portland, Oregon, area. At the southern limit of its distribution, depending upon the authority consulted, *caurinus* either existed allopatrically with the Western Crow *(hesperis)* or, according to some investigators, intergraded with this latter form. The majority opinion was that the two crows could be found together at the southern limit of the range, along the Olympic Peninsula, and in certain portions of the Puget Sound area.

Several prominent ornithologists (Brooks, 1942; Bowles, 1900: 84-85) argued forcefully for the recognition of *caurinus* as a distinct species on the basis of physical and ecological differences when compared with *hesperis*. Bowles regarded *caurinus* as being smaller, tamer, and a colonial nester. Brooks, conceding the fact that measurements between these two types of crows might overlap, characterized *caurinus* as having a different voice and a faster wing beat, and stated positively that the two were sympatric in the Sumas Prairie region of British Columbia, but they did not interbreed. For British Columbia in its entirety, Brooks and Swarth (1925: 80-81), Munro and Cowan (1947: 162-63), and Dickinson (1953: 169) generally regarded the (Western) Common Crow as an upland or inland species and the Northwestern Crow as a tidewater or coastal species, mostly confined to the beaches but occurring slightly inland along some of the major rivers. Thus, *caurinus* was principally characterized by having a different voice, smaller size, and occupying a habitat different from that of *hesperis*. The same general ideas as to differences from *hesperis* and habitat choice obtained for Washington (see Jewett, *et al.*, 1953). The crows from coastal Washington and the shores of Puget Sound were *caurinus*, but inland from these littorine areas were the larger Western Crows *(hesperis)*. On the University of Washington campus at Seattle, for example, Miller and Curtis (1940: 44) recorded both species, one presumably coming from the coastal strip of Puget Sound and the other from the more mountainous inland regions.

It is of considerable interest to note that no one could distinguish specimens of this supposed species, *caurinus*, from *C. b. hesperis* on the basis of color. Rather, morphological differences were presumably those of relative size (see, for example, Ridgway, 1904: 272), a series of characteristics which were of the same nature as those which constituted the *accepted* practice for distinguishing the various eastern subspecies of *C. brachyrhynchos*. But Brodkorb (in Blair, *et al.*, 1957: 534) presented a clear-cut distinction as follows: "tarsus 45-53 mm; culmen over 90% tarsus . . . *C. caurinus*. Tarsus 53-66.5 mm; culmen under 85% tarsus . . . *C. brachyrhynchos.*" Unfortunately no one bothered to consider the first-year birds, so that, actually, many of the very small specimens of *caurinus* were in fact first-year birds, obviously incorrectly comparable to adult specimens of *brachyrhynchos*.

My attention was first focused sharply on this situation while trying to identify various museum specimens from the Pacific Northwest. Here are some of the problems which were encountered: from the same locality, one specimen identified as *caurinus* and the other as *hesperis;* two specimens, virtually identical in all measurements, one labeled one species, the other, a different species; specimens from several localities labeled as either hybrids or intergrades; specimens which I could not personally identify (on the basis of measurements given in the literature) because they were too large or too small for a given type; the conflicting reports in the literature as to distribution and size of *caurinus;* reports that *hesperis* was increasing on Vancouver Island (Pearse, 1946: 7), whereas formerly it had been absent from this large island. This maze of problems was convincing evidence that specimen study alone would not resolve these conflicts, but rather, an intensive field study was necessary.

It was with this background information that I spent more than two months observing, collecting, and studying crows in Washington in the spring of 1958. During this time hundreds of field observations were transcribed, voices were carefully noted, specimens were taken, and 7,000 miles were traveled from southwestern to northwestern Washington, principally west of the Cascade Range, and in British Columbia. Field work centered in King and Snohomish counties around Seattle, but excursions of shorter duration were taken as far north as the San Juan Islands (Friday Harbor), Vancouver, B. C., the Skagit River valley, east to Snoqualmie Pass in the Cascade Range, Lake Wenatchee, Cle Elum, Mount Rainier, south to Grays Harbor, and west to the coastal areas around the Olympic Peninsula. A broad coverage of this critical area, involving collecting and observing, was considered to be the most effective procedure toward analyzing these crow populations and a resolution of the *caurinus* problems previously mentioned.

The results of these observations, collections, and examinations of museum specimens can best be presented by considering the characteristics of *caurinus* point by point, and then drawing conclusions from all the available evidence.

1. *Voice*. Anyone who has had experience with Common Crows in the East would likely be impressed by the harsher call notes of the Northwestern Crow, especially the usual calls of birds from Puget Sound and northward. From the outset, in order to provide a working basis, I tentatively assumed that the crows from coastal areas around Seattle were typical *caurinus* (as most authors insisted—see Jewett, *et al.*, 1953), and made detailed descriptions of their calls. My first impression (at Edmonds, Washington on April 2, 1958) while watching crows feeding along the beach, was that the call was similar to that of the Common Crow but was a harsher, lower-pitched *grar-r-r*. This call did not suggest to me the call of the eastern Fish Crow with which I have had considerable experience. Hellmayr (1934: 5), however, classified the Northwestern Crow as *Corvus ossifragus caurinus* Baird on the basis of the following statement: "Habits and call-note are described as being very different [from the Common Crow] and more like those of the Fish Crow. . . . Notwithstanding its widely separated range, *C. caurinus* has so much in common with the Fish Crow that Meise's proposition to link it with *C. ossifragus* seems to be the most satisfactory arrangement." As will be pointed out later, the usual call of *ossi-*

fragus is a terse, nasal *car* or *kark*, whereas that of *caurinus* is a more drawn-out, harsh, low-pitched *grar-r*.

Assuming that the usual calls of *caurinus* were distinctive on the shores of Puget Sound, my next step was to seek out *brachyrhynchos* inland, attempting to find the two forms together or even close to one another, principally to compare their voices. To my astonishment, however, the usual crow voices were of the harsh *grar-r* type to the foot of the Cascades, many miles up the river valleys (Skagit River, for example), and up to many of the mountain passes (Snoqualmie Pass, elevation 3010 feet). This suggested, at least at first, that *caurinus* might not be restricted to the immediate environs of tidal areas such as Puget Sound, but rather ranged inland at least into the Cascade Range.

More intensive observation of the call notes began to shed some further light on the problem. On April 11, 1958, at Edmonds near Seattle, the following observation was recorded: "Two crows returned to north slope ridge, first one then the other. One first perched high on a dead stub and gave 5 *grar*'s. There were some slight variations in these notes. Some were less harsh. . . . As they called, again it was apparent that one bird could give a lower, drawn-out *grar-r-r* and then a little later, a high-pitched, more rapid *gaw*. Usually the higher-pitched notes were in threes, whereas the lower ones were in fours or fives. One perched in a tree right over my head (50 ft.) and gave *grar*'s in twos—slow. Then it suddenly flew down to the beach, calling. These calls were more rapid, and higher-pitched almost like Common Crow." In the same general area on April 18, 1958, I recorded the following: "Two crows on south slope, one calling loudly. Whereas it had a certain degree of huskiness, at a distance the bird sounded like a Western Crow. Apparently, this one individual (and probably others) can give a high-pitched rapid *gaw* and a lower-pitched, slower *grar-r-r*, so that one moment you think it's *caurinus* and the next *hesperis*." Similarly, on April 25, 1958, "one crow returned to the top of the north slope where it perched high on a dead snag. Here it . . . called continuously. The notes were all of a hoarse or harsh quality, like a bird with a sore throat. Usually, they were slowly uttered =*grar-r-r*, *grar-r-r*. But on many occasions, this bird called more rapidly, at which time the notes were higher-pitched, and sounded like notes of the West. Crow." Since these variations were definitely known to be emitted by a single bird on several occasions, they cannot be attributed to sex or age. From these observations, it was apparent that a single bird could sound like either *caurinus* or *hesperis* depending upon the speed of delivery and perhaps other factors.

Subsequently, on several other occasions, I detected from a flock of crows both low-pitched and high-pitched notes, the one being of the harsh *grar-r* type and the other, of the near-*caw* type, though still atypical of Western (Common) Crows. Such observations were made not only near Seattle but also as far west as the ocean beaches of the Olympic Peninsula and as far south as Orting and Yelm (Pierce County).

These data suggested that the crows of mid- and southwestern Washington were not of a "pure stock" with respect to their voice characteristics, but if this hypothesis were true, then there should be some localities where voices were "pure." Field trips were subsequently planned to more distant localities in Washington to test this latter possibility. On May 15 and 16, 1958, intensive observations were carried out at Friday Harbor in the San Juan Islands, Washington: "It was significant to note the calls of all the crows [on this island]. At no time did I detect the crows giving anything but the harsh, low-pitched *grar-grar* notes." Although these observations suggested strongly (especially when used in conjunction with other mensural data presented below) that the birds of northwestern Washington were probably of the pure *caurinus* voice, the records of Pearse (1946) from Vancouver Island indicate that the voice variations, such as those noted in midwestern Washington, might occur there too.

Voices which were considered to be of pure *hesperis* type were detected in Washington on the eastern slopes of the Cascade Range at Lake Wenatchee and Plain, and at Morton, in Lewis County. (Specimens from these populations will be discussed below.)

In summary, these field observations suggest the following with respect to crow voices in the Pacific Northwest:
  a. Birds of extreme mid-southwestern Washington and those east of the Cascades are of the *hesperis* type.
  b. Birds of the Olympic Peninsula, mid- and probably much of northwestern Washington are of intermediate voice types, sometimes resembling *caurinus* and sometimes *hesperis*.
  c. Birds of the San Juan Islands, and perhaps from this locality northward, are of the *caurinus* type.

No doubt the most significant observation at this point was that of birds capable of rendering notes at least superficially resembling either kind of crow or sometimes not typical of either. Admittedly, these field observations were not as objective as one might desire; in the future, tape recordings and audiospectrographic analyses might be more instructive for further proof of this situation. Nonetheless, the present hypothesis of an extensive zone of intergradation helps to explain the frequent, though equivocal, observation of "both species" being in the same flock or geographic area.

2. *Habitat.* It is perfectly true, as dozens of authors have mentioned, that the Northwestern Crow of Alaska and British Columbia is essentially a coastal type, virtually never found any distance from this habitat. It is also probable that the mountain ranges and their dense forests along that portion of the Pacific coast present a formidable barrier toward much ingress by crows inland. In western Washington, however, especially around much of Puget Sound, mountain ranges are farther removed from the shores, thus leaving a comparatively wider belt of land, much of which has been opened into agricultural farmland. The same situation prevails from the southern part of Puget Sound southward to the Columbia River. In other words, the coastal habitat is immediately adjacent to some open farmland in this part of Washington, and the only habitat or geographic barrier is that of the Cascade Range and its forests in Central Washington.

On the Olympic Peninsula, except for the north central and northeastern parts, one usually finds dense forests and/or mountains close to the coasts, the result being the virtual absence of good crow habitat inland to any extent.

Between Puget Sound and the Cascade Range, a distance of some 50 or 60 miles, there is a band 20 or 30 miles in width of land which has been at least partially cleared for various kinds of farming practices. This band, adjacent to the coast, provides a habitat favorable to crows—open fields with nearby woods, as well as the very attractive pig farms and small-town garbage dumps. In the foothills of the Cascades where the forests are largely unbroken, crows are rather scarce, and are almost entirely restricted to river valleys and winding, mountain highways along which there is a limited amount of food available to crows.

As pointed out above, the crows found in this belt, from the shores of Puget Sound to the western slope of the Cascades, had some of the voice characteristics of *caurinus*. In southwestern Washington, where the upland agricultural habitats are continuous with those much farther south, crow voices were of an intermediate type, and it is here that the *caurinus* type, so typical of the more northern coastal strip intergrades with the *hesperis* type from the south.

3. *Measurements.* For the sake of convenience in analyzing measurements of these northwestern populations, specimens were divided into groups: Alaska; coastal British Columbia; northwestern Washington (north of King County which was the approximate halfway mark in the state); southwestern Washington (King County and southward and the Olympic Peninsula); interior Washington (east of the Cascades) along with interior British Columbia and Idaho; Oregon; and California. The number of adult specimens from these areas was, for the most part, adequate for statistical analyses. Graphic expressions of these measurements are presented in Figures 2 and 3.

Careful consideration of these several populations reveals the existence of significant mensural clines from California to Alaska. In adult males significant differences (lack of overlap in measurements) may be found for wings and tails between coastal British Columbia and northwestern Washington, and for the tarsus between the northwestern and southwestern Washington samples. In adult females, "breaks" are less clear due to a small sample from northwestern Washington. In most measurements of both sexes, however, the specimens from southwestern Washington are not significantly different from those of California and Oregon or from northwestern Washington (except possibly the tarsus). On the other hand, the specimens from interior Washington, British Columbia, and Idaho are definitely closer to the Oregon and California samples in their standard measurements.

Since tarsal length has been used in the past as a critical measurement in defining the species *caurinus*, let us examine the data more closely in Figures 2 and 3. Whereas in adult males, the two western Washington samples are significantly different from each other, the same difference cannot be demonstrated in adult females, even though the small sample size is a hindrance in this latter regard. Even so, the two western Washington samples are more nearly alike than either is to the interior Washington sample.

Nonetheless, examination of all these measurements shows that the northwestern crow populations are simply extremes—at the ends of clines—when compared with California and Oregon crows. This is further supportive evidence for the existence of a zone of intergradation in western Washington where the measurements are more or less intermediate between Alaska and California birds. It is also quite instructive to indicate here some critical intermediate specimens taken from the Cascade Range and certain localities in British Columbia. In these specimens, attention should be focused principally upon tarsal length, since this characteristic has been used most commonly in the past to separate *caurinus* from *hesperis*.

sex	age	locality	wing	tail	tarsus	bill
♂	im.	W. King Co., Wash.	291	172	53.4	32.9
♂	im.	W. King Co., Wash.	280	153	57.3	35.4
♀	im.	W. King Co., Wash.	276	—	50.3	30.5
♀	im.	W. King Co., Wash.	290	166	50.1	29.8
♀	ad.	Snoqualmie Pass, E. King Co., Wash.	—	—	49.9	28.1
♂	ad.	Chilliwack, B. C.	311	168	55.4	35.3
♂	ad.	Huntingdon, B. C.	294	166.5	52.9	34.6
♂	im.	Pemberton, B. C.	—	—	55.5	—
♂	im.	Sardis, B. C.	—	—	51.5	33.4
♂	ad.	Bella Coola, B. C.	295	163	50.7	38.2
♂	ad.	Bella Coola, B. C.	298	158.5	54.9	34.1
♀	ad.	Bella Coola, B. C.	276	149	49.7	32.1
♂	im.	Bella Coola, B. C.	288	—	55.7	34.5
♂	im.	Sumas Prairie, B. C.	288	163	53.4	34.2
♂	ad.	Agassiz, B. C.	312.5	175.5	52.2	35.6

All of the above localities are more or less intermediate between the coastal habitat and inland, mountainous areas. Snoqualmie Pass, Washington, is about twenty miles northwest of Cle Elum where a specimen typical of *C. b. hesperis* (measurements and voice) was taken. Many of the British Columbia localities mentioned above are in the Fraser River valley where evidently coastal forms of the crow interbreed with crows typical of interior British Columbia. The reader will recall that Brooks specifically designated the Sumas Prairie region as one where the two types of crows existed sympatrically, but the evidence introduced here strongly suggests that this is a region of intergradation with random interbreeding.

It is desirable to discuss here, at least briefly, specimens which various taxonomists have mislabeled as "*C. caurinus*" or as hybrids. Several adult specimens from northern, coastal Oregon identified as *caurinus*, had tarsi of 53.9 (♂), 52.9 (♀), and 52.1 (♀); these measurements are certainly not characteristic of specimens originally designated as *caurinus* on the basis of tarsal length. Others from the same region were first-year birds and do not differ significantly from other first-year specimens from Oregon; the reason why they were labeled *caurinus* was due, no doubt, to the taxonomist's failure to recognize that first-year birds are smaller than and cannot be compared with adults. Museum specimens marked *"caurinus* X *hesperis"* were taken at Tenino, Lapush, and Granville, Washington. The Granville specimen was a first-year male with a tarsus of 51.4 mm.; it had probably been out of the nest less than two months! The other specimens were adult females (tarsus, 54.0 and 53.4) and an adult male (54.1). These specimens, according to the present interpretation, would be considered intermediates from the zone of intergradation as previously outlined for southwestern Washington.

4. *Nesting*. Most authorities have agreed that the nests and eggs of the Northwestern and Western crows do not differ, except possibly in relative size. Others have claimed that different materials may be used in nesting, but this does not seem to be the case. Crow nests which were found in the vicinity of Seattle did not differ significantly from those which were studied from other parts of the country in various museums.

Colonial nesting, a feature especially attributed to *caurinus* by several authors, is a habit, although usually not well developed, found in many other crow populations. No doubt the crows of Alaska are conspicuous colonial nesters, but this same habit has been recorded sparingly in other portions of North America. Of signal interest here is the report of Emlen (1942) of colonial nesting in California, and if more were known about nesting crows in Oregon and Washington, I would predict that this behavior would be not uncommon in those areas too.

5. *Fossils*. The report of *Corvus caurinus* as a fossil in southern California (see Wetmore, 1956: 93) might be variously interpreted. It might mean that the Pleistocene crows of that region were as small as birds which now occur in coastal British Columbia and Alaska. It might also mean that the small tarsi, upon which this identification depended, were really from small, first-year birds typical of *C. brachyrhynchos hesperis*. In Table 1, the smallest tarsus of first-year females is 50.0 mm.; should a small specimen of this nature be encountered in fossil material without a large comparative sample, based upon sex and age groups, it might be incorrectly interpreted.

6. *Behavior*. Bowles (1900: 84), Kitchin (1949: 178), and other authors have called attention to the apparent tameness of crows in the Pacific Northwest. In fact, Bowles suggested that tameness was a distinctive feature of *caurinus*, whereas the Western Crow was presumably much more "wild." From Alaska south to the Olympic Peninsula crows may be found at times walking around in Indian villages or on lawns of urban areas, these birds evidencing little or no fear of the human inhabitants. This trait is obviously and strikingly different from the usual wary nature of Common Crows in most other parts of the continent. On the other hand, Dwight (1893: 10) noted that the Common Crows on Prince Edward Island were both abundant and tame.

It is probable that the tameness of certain crow populations is inversely proportional to the amount of human persecution of the species. The Indians of the Pacific Northwest, for example, rarely molest crows in their villages, but, rather, regard them as sacred birds. Conversely, in the midwestern and eastern United States where crows are heavily hunted especially for "sport," the birds are considered to be among the most wary and sagacious of all birds. Thus, the fact that crows of the Pacific Northwest are more tame than some other populations is closely correlated with the reduced hunting pressure on this segment of the population.

All the data presented here—voice, habitat choice, measurements—clearly negate any hypothesis that crows of northwestern North America represent a distinct species. Rather, the evidence points to the existence of a zone of intergradation in southwestern Washington and restricted localities in British Columbia where crows from the northern and southern populations freely interbreed with one another. In the absence of clear-cut differences between the northern and southern populations and the absence of reproductive isolation, it follows that the Northwestern Crow is simply a well-marked, ecologic subspecies of *C. brachyrhynchos*. It is with this conclusion in mind that the crows of Alaska and coastal British Columbia should be called *C. brachyrhynchos caurinus*, those of Washington west of the Cascade Range are intermediates, and those of Oregon, California, interior Washington and British Columbia, and Idaho, *C. b. hesperis*.

An interesting comparison might be suggested here with the ecologic races of song sparrows of San Francisco Bay (Marshall, 1948). In the case of this fringillid, isolation was effected by habitat distribution and open water, such that recognizable sub-

Common Crow

species of song sparrows could be associated with a given habitat. The crows of the Pacific Northwest, however, are not so sharply delimited to a specific habitat as was previously believed. The so-called "coastal *caurinus*" apparently occupies more xeric inland habitats when these are available and more or less adjacent to the littorine habitat. This being the case, only the extreme examples of these crows from British Columbia or Alaska exhibit the combined ecologic and morphologic characters which make them distinctive when compared with more southerly populations.

The type specimens, or four cotypes, of *caurinus* in the U. S. National Museum were all taken between 1854 and 1857 in the Puget Sound region. (In the American Museum of Natural History there is an immature specimen (42372) taken at Ft. Steilacoom, Washington, on April 25, 1856, by Dr. George Suckley. Since it was taken at the same time and place and by the same collector as one of the cotypes in the U. S. National Museum, it, too, should be recognized as a cotype.) Unfortunately, none of these five specimens is sexed, all are first-year birds except one, and only three were taken during the possible breeding season. It is of considerable interest to note that the measurements of these birds indicate that they represent the intergrades of that particular region, and are certainly atypical of *caurinus* as conceived in this study. It would seem that a new type locality should be fixed, one which is more typical of this coastal, small race; Alaska would be the preferred choice of localities.

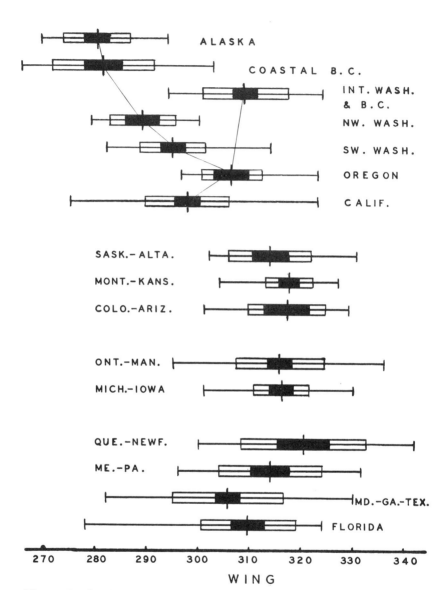

Figure 2. Graphic expressions of measurements in millimeters of adult male Corvus brachyrhynchos. Horizontal lines represent extremes; heavy vertical lines, the means; black rectangles, ± twice the standard error of the mean; open rectangles, ± one standard deviation.

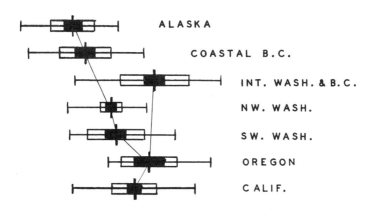
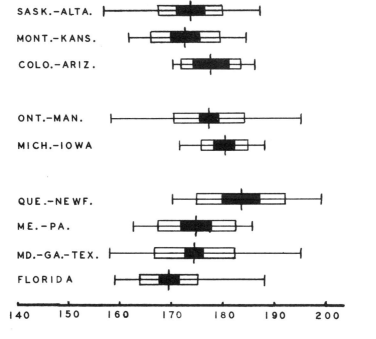

TAIL

Figure 2 (continued)

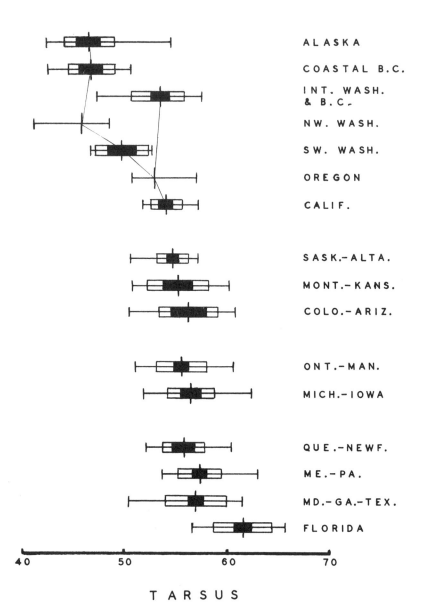

Figure 2 (continued)

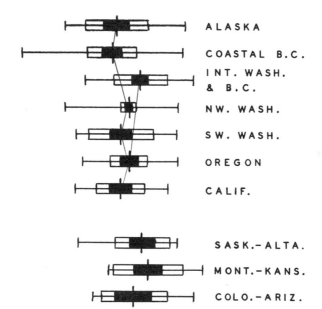
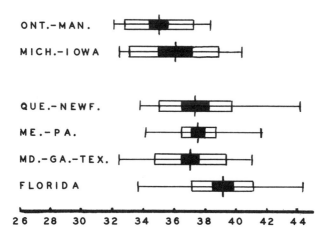

BILL

Figure 2 (continued)

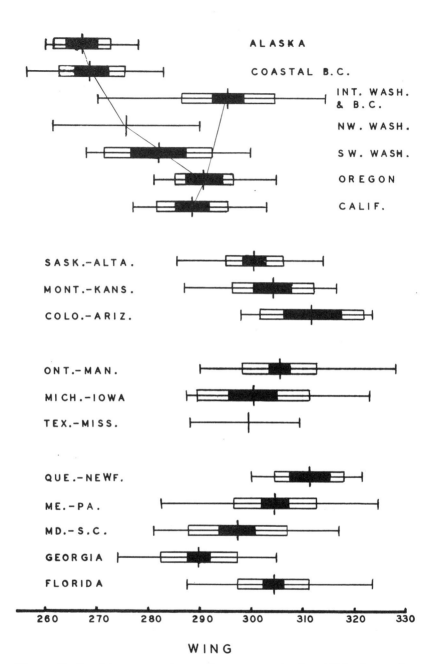

Figure 3. Graphic expressions of measurements in millimeters of adult female *Corvus brachyrhynchos*. Horizontal lines represent extremes; heavy vertical lines, the means; black rectangles, ± twice the standard error of the mean; open rectangle, ± one standard deviation.

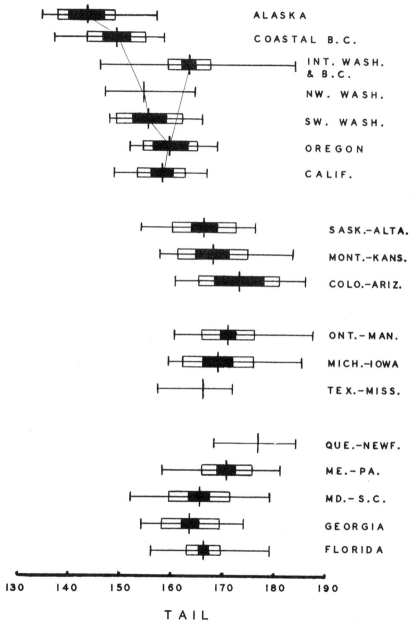

Figure 3 (continued)

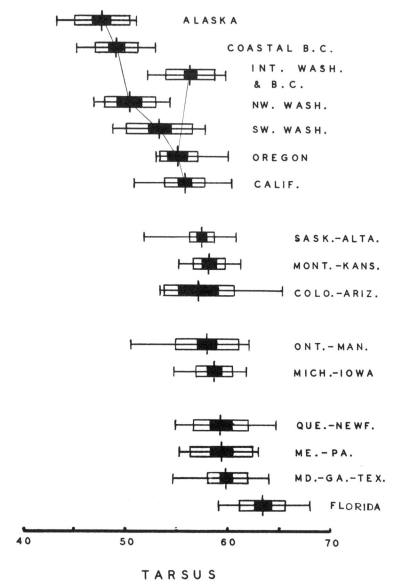

TARSUS

Figure 3 (continued)

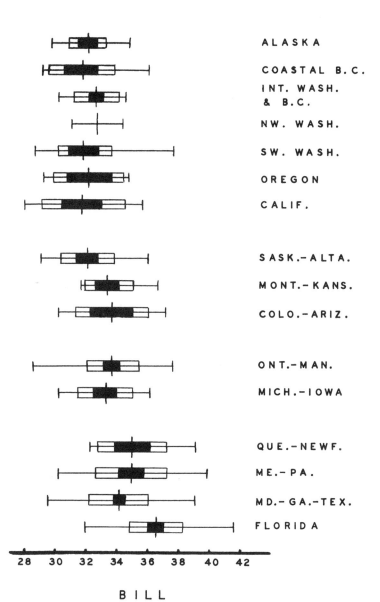

Figure 3 (continued)

TABLE 1

MEASUREMENTS OF CORVUS BRACHYRHYNCHOS

WING

	Adult Males				
	N	Mean	S.E.	Range	S.D.	C.V.
Que., N. B., N. S., Pr. Ed. Is., Newf.	25	320.4	2.43	300.0-342.0	12.17	3.8
Me., R. I., Conn., Mass., N. H., Vt., N. Y., N. J., Pa.	27	314.0	1.92	296.0-331.5	9.99	3.2
Md., D. C., Va., W. Va., Tenn., Ky., S. C., N. C.	33	308.3	1.95	283.5-330.0	11.19	3.6
Ga.	37	303.3	1.69	282.0-318.5	10.3	3.4
Fla.	32	309.7	1.63	278.0-324.0	9.25	3.0
Ont. and Man.	50	315.8	1.21	295.0-336.0	8.57	2.7
Mich., Ohio, Ill., Ind., Wisc., Iowa, Minn.	23	316.1	1.13	301.0-330.0	5.43	1.7
Tex., La., Ala., Ark., Miss.	11	306.4	3.01	296.5-321.0	10.0	3.3
Sask., Alta., Yukon, Keewatin	22	313.9	1.73	302.0-330.5	8.13	2.6
Mont., Wyo., N. D., S. D., Kans.	22	317.5	0.96	304.0-327.0	4.48	1.4
Colo., Okla., N. M., Ariz.	12	317.1	2.21	301.0-329.0	7.64	2.4
California	41	297.7	1.28	275.0-323.0	8.18	2.7
Oregon	12	306.3	1.71	296.5-323.0	5.93	1.9
West. Wash., King Co. southward	31	294.8	1.14	282.0-314.0	6.36	2.2
West. Wash., north of King Co.	15	288.9	1.68	279.0-300.0	6.51	2.3
Interior Wash., int. B. C., and Idaho	48	308.9	1.20	294.0-324.0	8.29	2.7
Coastal B. C.	29	281.3	1.85	265.0-303.0	9.95	3.5
Alaska	27	280.1	1.26	269.5-294.0	6.52	2.3

TAIL

Que., N. B., N. S., Pr. Ed. Is., Newf.	23	183.3	1.80	170.0-199.0	8.63	4.7

TABLE 1 (continued)

MEASUREMENTS OF CORVUS BRACHYRHYNCHOS

Adult Males

TAIL (cont.)	N	Mean	S.E.	Range	S.D.	C.V.
Me., R. I., Conn., Mass., N. H., Vt., N. Y., N. J., Pa.	26	174.8	1.49	162.5-185.5	7.59	4.3
Md., D. C., Va., W. Va., Tenn., Ky., S. C., N. C.	35	174.5	1.46	159.0-195.0	8.62	4.9
Georgia	40	174.4	0.87	158.0-189.0	5.51	3.2
Florida	32	169.4	1.00	159.0-188.0	5.67	3.3
Ont. and Man.	51	177.1	0.97	158.0-195.0	6.90	3.9
Mich., Ohio, Ill., Ind., Wisc., Iowa, Minn.	22	180.1	1.01	171.5-188.0	4.74	2.6
Tex., La., Ala., Ark., Miss.	11	174.4	1.05	170.5-188.0	3.47	2.0
Sask., Alta., Yukon, Keewatin	22	173.5	1.37	156.5-187.0	6.42	3.7
Mont., Wyo., N. D., S. D., Kans.	24	172.6	1.37	161.5-184.5	6.72	3.9
Colo., Okla., N. M., Ariz.	11	177.5	1.77	170.0-186.0	5.88	3.3
California	42	165.1	0.69	153.0-177.0	4.49	2.7
Oregon	14	168.0	1.48	160.0-180.0	5.55	3.3
West Wash., King Co. southward	31	161.5	1.00	152.5-173.0	5.58	3.5
West. Wash., north of King Co.	14	160.6	0.55	152.0-167.5	2.05	1.3
Interior Wash., int. B. C., and Idaho	49	169.0	0.96	153.5-182.0	6.75	4.0
Coastal B. C.	28	155.6	0.96	144.5-167.0	5.08	3.3
Alaska	22	153.1	0.89	143.0-162.0	4.21	2.7
TARSUS						
Que., N. B., N. S., Pr. Ed. Is., Newf.	26	59.3	0.54	54.9-64.8	2.74	4.6
Me., R. I., Conn., Mass., N. H., Vt., N. Y., N. J., Pa.	31	59.4	0.55	55.2-63.0	3.09	5.2

TABLE 1 (continued)

MEASUREMENTS OF CORVUS BRACHYRHYNCHOS

Adult Males

TARSUS (cont.)	N	Mean	S.E.	Range	S.D.	C.V.
Md., D. C., Va., W. Va., Tenn., Ky., S. C., N. C.	33	59.1	0.23	54.8-63.0	1.34	2.3
Georgia	40	60.4	0.42	55.0-64.0	2.68	4.4
Florida	36	63.4	0.38	59.1-68.0	2.29	3.6
Ont. and Man.	51	58.0	0.43	50.6-62.2	3.10	5.3
Mich., Ohio, Ill., Ind., Wisc., Iowa, Minn.	25	58.7	0.35	54.8-61.9	1.77	3.0
Tex., La., Ala., Ark., Miss.	11	59.7	0.56	56.8-63.6	1.85	3.1
Sask., Alta., Yukon, Keewatin	23	57.5	0.26	51.8-60.9	1.24	2.2
Mont., Wyo., N. D., S. D., Kans.	22	58.2	0.35	55.2-61.3	1.63	2.8
Colo., Okla., N. M., Ariz.	12	57.2	1.01	53.3-65.3	3.51	6.1
California	45	55.9	0.29	50.9-60.5	1.97	3.5
Oregon	16	55.2	0.49	53.1-60.1	1.95	3.5
West. Wash., King Co. southward	34	53.4	0.55	48.8-57.9	3.22	6.0
West. Wash., north of King Co.	16	50.5	0.62	47.0-54.0	2.47	4.9
Interior Wash., int. B. C., and Idaho	49	56.4	0.34	52.3-59.9	2.37	4.2
Coastal B. C.	31	49.2	0.38	45.3-53.0	2.14	4.3
Alaska	36	47.8	0.45	43.4-51.2	2.72	5.7
BILL						
Que., N. B., N. S., Pr. Ed. Is., Newf.	26	37.3	0.47	33.8-44.1	2.40	6.4
Me., R. I., Conn., Mass., N. H., Vt., N. Y., N. J., Pa.	32	37.5	0.20	34.1-41.5	1.15	3.1
Md., D. C., Va., W. Va., Tenn., Ky., S. C., N. C.	31	36.7	0.17	32.4-39.6	0.95	2.6

TABLE 1 (continued)

MEASUREMENTS OF CORVUS BRACHYRHYNCHOS

BILL (cont.)	Adult Males					
	N	Mean	S.E.	Range	S.D.	C.V.
Georgia	40	37.3	0.23	34.0-41.0	1.48	4.0
Florida	34	39.1	0.34	33.6-44.3	1.96	5.0
Ont. and Man.	52	34.9	0.32	32.0-38.2	2.32	6.6
Mich., Ohio, Ill., Ind., Wisc., Iowa, Minn.	24	36.0	0.58	33.7-40.3	2.86	7.9
Tex., La., Ala., Ark., Miss.	11	37.1	0.56	35.1-39.2	1.85	4.9
Sask., Alta., Yukon, Keewatin	22	34.6	0.39	30.6-36.9	1.84	5.3
Mont., Wyo., N. D., S. D., Kans.	24	35.0	0.46	32.5-38.5	2.27	6.5
Colo., Okla., N. M., Ariz.	12	34.1	0.62	31.5-38.0	2.15	6.3
California	45	33.2	0.34	30.3-36.3	1.62	4.9
Oregon	15	33.7	0.32	30.8-36.9	1.23	3.6
West. Wash., King Co. southward	35	33.2	0.36	30.3-36.8	2.11	6.4
West. Wash., north of King Co.	15	33.7	0.10	29.6-36.9	0.40	1.2
Interior Wash., int. B. C., and Idaho	49	34.4	0.26	30.9-37.9	1.81	5.3
Coastal B. C.	30	32.6	0.30	26.8-37.2	1.63	5.0
Alaska	27	32.9	0.39	29.6-37.3	2.01	6.1

WING	Adult Females					
Que., N. B., N. S., Pr. Ed. Is., Newf.	12	311.3	1.95	300.0-321.5	6.74	2.2
Me., R. I., Conn., Mass., N. H., Vt., N. Y., N. J., Pa.	36	304.7	1.35	282.5-324.5	8.08	2.7
Md., D. C., Va., W. Va., Tenn., Ky., S. C., N. C.	31	297.4	1.72	281.0-317.0	9.60	3.2
Georgia	48	289.9	1.09	274.0-305.0	7.58	2.6
Florida	49	304.2	1.00	287.5-323.5	7.00	2.3

TABLE 1 (continued)

MEASUREMENTS OF CORVUS BRACHYRHYNCHOS

WING (cont.)

Adult Females

	N	Mean	S.E.	Range	S.D.	C.V.
Ont. and Man.	47	305.5	1.06	290.0-328.0	7.26	2.4
Mich., Ohio, Ill., Ind., Wisc., Iowa, Minn.	22	300.4	2.32	287.5-323.0	10.90	3.6
Tex., La., Ala., Ark., Miss.	8	299.4		288.0-309.5		
Sask., Alta., Yukon, Keewatin	25	300.6	1.13	285.5-314.0	5.63	1.9
Mont., Wyo., N. D., S. D., Kans.	18	304.2	1.92	287.0-316.5	8.12	2.7
Colo., Okla., N. M., Ariz.	12	311.9	2.88	298.0-323.5	9.96	3.2
California	17	288.5	1.72	277.0-303.0	7.09	2.5
Oregon	10	290.9	1.87	281.0-305.0	5.92	2.0
West. Wash., King Co. southward	15	282.0	2.69	267.0-300.0	10.40	3.7
West. Wash., north of King Co.	5	275.9		261.5-290.0		
Interior Wash., int. B. C., and Idaho	37	295.6	1.52	270.0-314.5	9.22	3.1
Coastal B. C.	17	268.9	1.58	256.5-283.0	6.52	2.4
Alaska	14	267.1	1.52	260.0-278.0	5.69	2.1

TAIL

	N	Mean	S.E.	Range	S.D.	C.V.
Que., N. B., N. S., Pr. Ed. Is., Newf.	8	176.8		168.0-184.0		
Me., R. I., Conn., Mass., N. H., Vt., N. Y., N. J., Pa.	32	170.6	0.84	158.0-181.0	4.78	2.8
Md., D. C., Va., W. Va., Tenn., Ky., S. C., N. C.	31	165.3	1.05	152.0-179.0	5.84	3.5
Georgia	48	163.6	0.81	154.0-174.0	5.59	3.4
Florida	48	166.2	0.47	156.0-179.0	3.25	2.0
Ont. and Man.	49	170.5	0.76	160.0-187.0	5.34	3.1
Mich., Ohio, Ill., Ind., Wisc., Iowa, Minn.	22	168.7	1.46	159.0-185.0	6.83	4.0

TABLE 1 (continued)

MEASUREMENTS OF CORVUS BRACHYRHYNCHOS

Adult Females

TAIL (cont.)	N	Mean	S.E.	Range	S.D.	C.V.
Tex., La., Ala., Ark., Miss.	9	165.9		157.0-171.5		
Sask., Alta., Yukon, Keewatin	25	165.6	1.26	153.5-175.5	6.28	3.8
Mont., Wyo., N. D., S. D., Kans.	18	167.2	1.61	157.0-183.0	6.81	4.1
Colo., Okla., N. M., Ariz.	11	172.5	2.34	160.0-185.5	7.78	4.5
California	18	157.3	1.08	148.0-166.0	4.56	2.9
Oregon	10	158.8	1.68	151.0-168.0	5.32	3.4
West. Wash., King Co. southward	15	154.7	1.67	147.0-165.0	6.48	4.2
West. Wash., north of King Co.	5	153.7		146.0-163.5		
Interior Wash., int. B. C., and Idaho	37	162.2	0.67	145.0-183.0	4.10	2.5
Coastal B. C.	17	148.1	1.38	136.0-157.5	5.69	3.8
Alaska	11	142.2	1.70	133.5-156.0	5.66	4.0
TARSUS						
Que., N. B., N. S., Pr. Ed. Is., Newf.	15	55.9	0.53	52.1-60.4	2.06	3.7
Me., R. I., Conn., Mass., N. H., Vt., N. Y., N. J., Pa.	35	57.3	0.37	53.7-63.0	2.17	3.8
Md., D. C., Va., W. Va., Tenn., Ky., S. C., N. C.	31	56.1	0.33	52.2-59.3	1.85	3.3
Georgia	48	57.3	0.38	50.4-61.5	2.65	4.6
Florida	49	61.5	0.41	56.6-65.7	2.90	4.7
Ont. and Man.	51	55.6	0.35	51.1-60.6	2.51	4.5
Mich., Ohio, Ill., Ind., Wisc., Iowa, Minn.	22	56.5	0.48	51.9-62.3	2.26	4.0
Tex., La., Ala., Ark., Miss.	10	58.6	0.56	55.4-61.5	1.77	3.0

TABLE 1 (continued)

MEASUREMENTS OF CORVUS BRACHYRHYNCHOS

Adult Females

TARSUS (cont.)	N	Mean	S.E.	Range	S.D.	C.V.
Sask., Alta., Yukon, Keewatin	26	54.8	0.32	51.6-57.2	1.64	2.9
Mont., Wyo., N. D., S. D., Kans.	18	55.3	0.71	50.9-60.2	3.02	5.5
Colo., Okla., N. M., Ariz.	12	56.3	0.85	50.5-60.8	2.94	5.2
California	19	54.1	0.36	51.9-57.3	1.55	2.9
Oregon	10	53.0	0.65	50.9-57.1	2.05	3.9
West. Wash., King Co. southward	15	49.9	0.70	46.9-52.8	2.69	5.4
West. Wash., north of King Co.	5	46.0		41.3-48.7		
Interior Wash., int. B. C., and Idaho	39	53.6	0.45	47.5-57.6	2.80	5.2
Coastal B. C.	17	46.9	0.55	41.7-50.8	2.25	4.8
Alaska	17	46.7	0.61	42.5-54.7	2.51	5.4
BILL						
Que., N. B., N. S., Pr. Ed. Is., Newf.	15	34.9	0.60	32.2-39.0	2.34	6.7
Me., R. I., Conn., Mass., N. H., Vt., N. Y., N. J., Pa.	32	34.9	0.40	30.1-39.7	2.28	6.5
Md., D. C., Va., W. Va., Tenn., Ky., S. C., N. C.	30	33.5	0.34	29.5-37.3	1.86	5.6
Georgia	47	34.3	0.08	31.0-39.0	0.57	1.7
Florida	46	36.5	0.25	31.9-41.5	1.69	4.6
Ont. and Man.	53	33.6	0.23	28.3-37.8	1.71	5.1
Mich., Ohio, Ill., Ind., Wisc., Iowa, Minn.	22	33.1	0.39	30.1-36.0	1.82	5.5
Tex., La., Ala., Ark., Miss.	10	35.2	0.37	33.7-37.7	1.17	3.3
Sask., Alta., Yukon, Keewatin	25	31.9	0.37	28.8-35.8	1.84	5.8

TABLE 1 (continued)

MEASUREMENTS OF CORVUS BRACHYRHYNCHOS

BILL (cont.)

Adult Females

	N	Mean	S.E.	Range	S.D.	C.V.
Mont., Wyo., N. D., S. D., Kans.	17	33.2	0.41	31.7-36.4	1.67	5.0
Colo., Okla., N. M., Ariz.	12	33.5	0.70	30.0-37.0	2.41	7.2
California	18	31.5	0.65	27.7-35.4	2.75	8.7
Oregon	10	31.9	0.74	28.9-34.5	2.33	7.3
West. Wash., King Co. southward	15	31.6	0.48	28.3-37.4	1.85	5.9
West. Wash., north of King Co.	4	32.4		30.7-34.1		
Interior Wash., int. B. C., and Idaho	38	32.3	0.24	29.8-34.3	1.46	4.5
Coastal B. C.	15	31.4	0.58	28.7-35.7	2.23	7.1
Alaska	17	31.7	0.29	29.3-34.5	1.18	3.7

WING

First-Year Males

	N	Mean	S.E.	Range	S.D.	C.V.
Me., R. I., Conn., Mass., N. H., Vt., N. Y., N. J., Pa.	20	301.7	2.25	280.0-328.5	10.04	3.3
Md., D. C., Va., W. Va., Tenn., Ky., S. C., N. C.	16	292.3	2.38	272.0-314.0	9.53	3.3
Georgia	24	289.2	1.34	271.0-306.0	6.58	2.3
Florida	12	293.5	3.60	277.5-316.0	12.50	4.3
Mich., Ohio., Ill., Ind., Wisc., Iowa, Minn.	11	299.4	2.15	282.0-308.0	7.14	2.4
Tex., La., Ala., Ark., Miss.	9	295.5		283.5-309.0		
Sask., Alta., Yukon, Keewatin	4	306.1		300.5-316.0		
Mont., Wyo., N. D., S. D., Kans.	11	302.0	1.86	288.0-311.0	6.16	2.0
California	9	290.1		274.5-302.0		
Oregon	3	299.2		285.0-308.0		

TABLE 1 (continued)

MEASUREMENTS OF CORVUS BRACHYRHYNCHOS

WING (cont.)	First-Year Males					
	N	Mean	S.E.	Range	S.D.	C.V.
West. Wash., King Co. southward	22	286.2	1.93	272.0-307.0	9.06	3.2
West. Wash., north of King Co.	3	287.3		277.0-293.0		
Interior Wash., int. B. C., and Idaho	12	296.9	1.59	292.0-306.0	5.50	1.9
Coastal B. C.	9	279.7		274.0-286.0		
TAIL						
Me., R. I., Conn., Mass., N. H., Vt., N. Y., N. J., Pa.	15	169.3	1.72	156.0-181.5	6.67	3.9
Md., D. C., Va., W. Va., Tenn., Ky., S. C., N. C.	17	162.5	1.55	152.5-173.5	6.37	3.9
Georgia	24	164.1	1.42	155.0-178.0	6.97	4.2
Florida	10	162.2	2.36	152.0-175.0	7.45	4.6
Mich., Ohio, Ill., Ind., Wisc., Iowa, Minn.	11	169.4	2.39	154.0-180.0	7.94	4.7
Tex., La., Ala., Ark., Miss.	7	164.8		156.0-172.0		
Sask., Alta., Yukon Keewatin	4	174.6		164.0-180.5		
Mont., Wyo., N. D., S. D., Kans.	11	168.5	2.08	156.5-181.0	6.92	4.1
California	7	161.8		151.5-173.0		
West. Wash., King Co. southward	16	162.3	1.68	152.0-172.0	6.70	4.1
West. Wash., north of King Co.	3	160.0		156.0-163.0		
Interior Wash., int. B. C., and Idaho	8	163.2		156.0-170.0		
Coastal B. C.	7	157.1		150.0-170.5		
TARSUS						
Me., R. I., Conn., Mass., N. H., Vt., N. Y., N. J., Pa.	24	58.3	0.43	51.6-62.0	2.11	3.6

TABLE 1 (continued)

MEASUREMENTS OF CORVUS BRACHYRHYNCHOS

TARSUS (cont.)

First-Year Males

	N	Mean	S.E.	Range	S.D.	C.V.
Md., D. C., Va., W. Va., Tenn., Ky., S. C., N. C.	19	58.1	0.29	55.5-63.1	1.27	2.2
Georgia	27	59.6	0.37	55.0-63.0	1.93	3.2
Florida	16	62.3	0.79	58.2-65.7	3.15	5.1
Ont. and Man.	4	57.2		53.6-59.9		
Mich., Ohio, Ill., Ind., Wisc., Iowa, Minn.	19	57.9	0.44	54.1-61.3	1.91	3.3
Tex., La., Ala., Ark., Miss.	10	58.9	1.01	53.1-65.5	3.19	5.4
Sask., Alta., Yukon, Keewatin	5	57.2		55.9-58.0		
Mont., Wyo., N. D., S. D., Kans.	11	58.0	0.50	55.4-59.8	1.67	2.9
Colo., Okla., N. M., Ariz.	3	53.9		52.3-56.0		
California	20	56.5	0.54	51.6-59.7	2.40	4.2
Oregon	4	54.7		51.5-58.4		
West. Wash., King Co. southward	32	53.0	0.50	48.4-57.1	2.82	5.3
West. Wash., north of King Co.	4	53.4		50.9-57.1		
Interior Wash., int. B. C., and Idaho	15	56.0	0.32	53.9-57.7	1.22	2.2
Coastal B. C.	30	49.3	0.32	45.1-53.0	1.75	3.5
Alaska	11	48.0	0.40	45.6-49.5	1.34	2.8

BILL

	N	Mean	S.E.	Range	S.D.	C.V.
Me., R. I., Conn., Mass., N. H., Vt., N. Y., N. J., Pa.	24	35.5	0.44	32.0-39.2	2.13	6.0
Md., D. C., Va., W. Va., Tenn., Ky., S. C., N. C.	17	35.7	0.51	31.8-38.4	2.09	5.9
Georgia	29	36.7	0.39	32.0-41.0	2.10	5.7
Florida	15	36.8	0.42	33.4-42.8	1.62	4.4

TABLE 1 (continued)
MEASUREMENTS OF CORVUS BRACHYRHYNCHOS

First-Year Males

BILL (cont.)	N	Mean	S.E.	Range	S.D.	C.V.
Ont. and Man.	3	33.7		28.3-37.1		
Mich., Ohio, Ill., Ind., Wisc., Iowa, Minn.	19	34.9	0.44	29.2-37.6	1.93	5.5
Tex., La., Ala., Ark., Miss.	10	37.5	0.46	35.0-40.5	1.46	3.9
Sask., Alta., Yukon, Keewatin	5	33.9		32.6-35.4		
Mont., Wyo., N. D., S. D., Kans.	13	33.6	0.65	29.6-37.0	2.36	7.0
Colo., Okla., N. M., Ariz.	3	32.2		31.9-32.3		
California	21	33.5	0.38	31.1-35.6	1.75	5.2
Oregon	4	32.6		30.1-34.1		
West. Wash., King Co. southward	29	33.2	0.40	30.3-37.6	2.14	6.4
West. Wash., north of King Co.	4	33.9		29.1-38.4		
Interior Wash., int. B. C., and Idaho	16	33.7	0.60	29.8-36.9	2.38	7.1
Coastal B. C.	27	32.6	0.40	27.6-34.6	2.08	6.4
Alaska	4	32.4		30.1-34.1		

First-Year Females

WING	N	Mean	S.E.	Range	S.D.	C.V.
Me., R. I., Conn., Mass., N. H., Vt., N. Y., N. J., Pa.	21	290.2	1.74	275.0-303.0	7.99	2.8
Md., D. C., Va., W. Va., Tenn., Ky., S. C., N. C.	11	286.1		276.0-300.0		
Georgia	25	279.7	1.31	265.0-298.0	6.55	2.3
Mich., Ohio, Ill., Ind., Wisc., Iowa, Minn.	7	291.0		269.0-305.0		
Tex., La., Ala., Ark., Miss.	7	276.8		256.0-293.0		
Sask., Alta., Yukon, Keewatin	4	288.9		285.0-293.5		

TABLE 1 (continued)

MEASUREMENTS OF CORVUS BRACHYRHYNCHOS

First-Year Females

WING (cont.)	N	Mean	S.E.	Range	S.D.	C.V.
Mont., Wyo., N. D., S. D., Kans.	7	294.5		286.5-303.0		
Colo., Okla., N. M., Ariz.	5	294.3		286.5-301.0		
California	7	282.9		267.0-298.0		
West. Wash., King Co. southward	15	272.7	2.04	253.0-290.0	7.88	2.9
West. Wash., north of King Co.	6	267.9		259.0-286.5		
Interior Wash., int. B. C., and Idaho	9	290.5		274.0-309.0		
Coastal B. C.	4	269.5		257.0-278.0		
Alaska	4	272.4		260.0-284.0		

TAIL

	N	Mean	S.E.	Range	S.D.	C.V.
Me., R. I., Conn., Mass., N. H., Vt., N. Y., N. J., Pa.	14	163.5	1.30	153.0-170.0	4.86	2.9
Md., D. C., Va., W. Va., Tenn., Ky., S. C., N. C.	8	159.3		151.5-169.5		
Georgia	23	158.0	1.70	140.0-168.0	8.13	5.1
Mich., Ohio, Ill., Ind., Wisc., Iowa, Minn.	6	161.0		141.5-170.0		
Sask., Alta., Yukon, Keewatin	3	161.2		159.0-162.5		
Mont., Wyo., N. D., S. D., Kans.	4	164.8		161.0-174.0		
Colo., Okla., N. M., Ariz.	5	162.6		156.5-169.0		
California	3	156.7		154.0-160.0		
West. Wash., King Co. southward	14	154.4	1.74	137.0-166.0	6.50	4.2
West. Wash., north of King Co.	5	149.0		142.0-158.0		
Interior Wash., int. B. C., and Idaho	4	158.3		146.0-164.5		

TABLE 1 (continued)

MEASUREMENTS OF CORVUS BRACHYRHYNCHOS

TAIL (cont.)	First-Year Females					
	N	Mean	S.E.	Range	S.D.	C.V.
Coastal B. C.	4	151.4		141.0-156.0		
Alaska	4	150.3		136.0-163.0		

TARSUS

	N	Mean	S.E.	Range	S.D.	C.V.
Me., R. I., Conn., Mass., N. H., Vt., N. Y., N. J., Pa.	31	56.9	0.29	53.0-61.3	1.59	2.8
Md., D. C., Va., W. Va., Tenn., Ky., S. C., N. C.	11	55.8		52.9-59.7		
Georgia	27	58.0	0.63	51.0-62.0	3.25	5.6
Florida	14	59.6	0.44	54.9-64.3	1.65	2.8
Ont. and Man.	3	54.8		53.3-57.5		
Mich., Ohio, Ill., Ind., Wisc., Iowa, Minn.	10	56.9	0.55	52.8-58.7	1.73	3.0
Tex., La., Ala., Ark., Miss.	7	54.9		46.3-61.2		
Sask., Alta., Yukon, Keewatin	7	55.3		52.7-58.9		
Mont., Wyo., N. D., S. D., Kans.	9	56.6		53.2-61.6		
Colo., Okla., N. M., Ariz.	7	56.0		54.7-57.7		
California	18	54.6	0.48	50.0-58.7	2.03	3.7
Oregon	3	52.7		51.8-53.9		
West. Wash., King Co. southward	19	49.8	0.48	43.0-55.4	2.10	4.2
West. Wash., north of King Co.	7	51.5		47.6-54.5		
Interior Wash., int. B. C., and Idaho	19	53.7	0.41	50.6-56.8	1.79	3.3
Coastal B. C.	13	48.0	0.75	44.7-50.8	2.72	5.7
Alaska	14	45.9	0.47	42.4-48.4	1.76	3.8

TABLE 1 (continued)

MEASUREMENTS OF CORVUS BRACHYRHYNCHOS

BILL

First-Year Females

	N	Mean	S.E.	Range	S.D.	C.V.
Me., R. I., Conn., Mass., N. H., Vt., N. Y., N. J., Pa.	29	33.8	0.43	29.5-38.2	2.32	6.7
Md., D. C., Va., W. Va., Tenn., Ky., S. C., N. C.	10	33.6		31.3-35.7		
Georgia	26	34.4	0.36	30.1-38.0	1.83	5.3
Florida	14	35.0	0.29	32.1-36.8	1.07	3.1
Ont. and Man.	3	33.2		32.4-33.8		
Mich., Ohio, Ill., Ind., Wisc., Iowa, Minn.	10	32.9	0.73	28.7-36.6	2.30	7.0
Tex., La., Ala., Ark., Miss.	6	34.3		32.1-36.8		
Sask., Alta., Yukon, Keewatin	7	31.7		30.8-33.2		
Mont., Wyo., N. D., S. D., Kans.	10	34.2		30.7-37.9		
Colo., Okla., N. M., Ariz.	7	33.1		30.0-34.7		
California	17	30.5		27.6-33.1		
Oregon	3	33.9		32.9-34.9		
West. Wash., King Co. southward	18	31.3	0.41	28.9-34.8	1.74	5.6
West. Wash., north of King Co.	7	31.5		30.1-34.4		
Interior Wash., int. B. C., and Idaho	19	32.0	0.50	28.2-34.2	2.20	6.9
Coastal B. C.	10	30.6	0.56	29.2-32.7	1.76	5.8
Alaska	6	32.8		31.1-34.7		

4. FISH CROW

Corvus ossifragus Wilson

Corvus ossifragus Wilson

Geographic Range
  Specimens examined, 171: from Louisiana, 6; Mississippi, 9; Georgia, 19; Alabama, 8; Florida, 72; North Carolina, 6; South Carolina, 8; Virginia, 7; Maryland, 1; Delaware, 1; New Jersey, 9; New York, 18; Pennsylvania, 4; District of Columbia, 1; and Connecticut, 2; others reported in the literature from Texas, Arkansas, Tennessee, Rhode Island, and Massachusetts (breeding?).

Throughout this range the Fish Crow reaches its peak of abundance in and is closely associated with the Coastal Plain (Figure 1). In recent years, however, records indicate that this species is expanding its geographic range inland into the Piedmont Plateau and northward. For Georgia there are recent Piedmont records at Athens (Johnston, 1947), Thompson (Denton, 1950), and along the shores of Lake Sinclair in upper Baldwin County. In western Arkansas and eastern Oklahoma, Wilhelm (in press) has reported range extensions in recent years, with nests being found along the Arkansas River near Ft. Smith, Arkansas, and Moffett, Oklahoma. Nice and Nice (1924) did not record this species from Oklahoma. On the Mississippi River, Ben Coffey (letter) reported: "Only in the last decade have we recorded the species [at Memphis and] . . . only in recent years have we seen the Fish Crow [at] . . . Reelfoot Lake."
  In addition to its common occurrence in the Coastal Plain of Maryland, the Fish Crow was recorded by Stewart and Robbins (1958: 221) as "uncommon and local in the Piedmont, and Ridge and Valley sections (occurring in Frederick and Hagerstown Valleys)."

For Rhode Island, Richard Bowen stated (letter) that Fish Crows were extremely rare before 1941, but near Warren the species has increased in numbers, the first nest for the state being discovered in 1943. In the following years, the species not only increased in the breeding season, but was also recorded as a winter straggler. After 1946, with the discovery of thirty Fish Crows at Bristol, in May, Bowen stated: "This was the beginning of a general spread of Fish Crows in the Upper Narraganset Bay area and from this time on the species became more common generally."

Of its occurrence in Massachusetts, Griscom and Snyder (1955: 168) wrote: "Formerly a straggler from the south, now occurring more frequently as it breeds regularly in Warren County, Rhode Island, and wanders over the border at Swansea, Seekonk, and Westport. . . . It is now occasionally reported in spring in southeastern Massachusetts along the coast from Cambridge south, and at Longmeadow in the Connecticut Valley. . . . The proved breeding of the Fish Crow in this state may be confidently expected." Since that time, James C. Greenway (letter) reported that two young Fish Crows were taken in July, 1957, in Hockamock Swamp, Mass., there being a good possibility that the birds were raised nearby.

Historically, it is of interest to note that Baird, Brewer, and Ridgway (1875: 252) believed that the Fish Crow did not occur north of New Jersey. Whether this statement reflected incomplete knowledge on the part of the authors or a statement of the facts is not known, but apparently there are no early breeding specimens to invalidate their contention.

Habitat. The older, standard works on the distribution of birds (for example, Howell, 1932) generally listed the Fish Crow as an inhabitant of ocean beaches, river valleys, and lake shores. The ecologic terms "littoral" and "riparian" rather generally express the habitat choice of this species.

In recent years, however, the species has not only extended its geographic range northward, as indicated above, but also has begun to occupy more xeric communities. Stewart and Robbins (1958: 221) described the habitat as "wood margin, field, shore, and marsh habitats that are adjacent to tidewater; in the interior, also occurs sparingly in Frederick and Washington Counties in agricultural fields and field borders." In central Georgia where the author has studied this species extensively, Fish Crows follow the Ocmulgee River northward in the spring to the Fall Line at Macon, and then fan out into the upland regions sometimes five or ten miles from the nearest river or lake. It is in this locality that I have taken the species in upland forests of *Pinus taeda* and *P. echinata,* in pecan orchards, and occasionally in dry, old fields sometimes in company with Common Crows. Similarly, crows were studied intensively in the summer of 1956 at the Savannah River Plant of the Atomic Energy Commission near Aiken, South Carolina, and, here, too, Fish Crows ranged widely into upland pine forests, fields, and roadsides.

It is precisely in this group of upland habitats that Fish and Common crows are sympatric in both a geographic and an ecologic sense, because, as indicated previously, the Common Crow is principally a species of agricultural farm lands. Careful study of the interspecific relationships in upland habitats has yielded some interesting facts. At the Savannah River Plant historical aspects of plant succession, when correlated with the distribution of crows, showed that Common Crows were more abundant in the first few years of the Plant's existence and gradually diminished in numbers as the abandoned farm lands reverted to natural fields of grasses, forbs, and young pines. It is evident that the Common Crows depended in large part upon their ability to obtain agricultural products and associated insects and as these foods disappeared the birds diminished in numbers. Fish Crows, on the other hand, apparently were formerly restricted to the immediate environs of the Savannah River, but, as human agricultural practices ceased and as the Common Crow diminished, the Fish Crow increased in numbers in the uplands. Evidently, the Fish Crows could thrive on native foods (berries and other fruits, seeds, and arthropods) as well as road kills (snakes, mammals, and amphibians).

Some of the southeastern coastal areas are more heavily populated by Fish Crows than Common Crows. For example, on the Georgia Coastal Islands which face the ocean and are largely uncultivated, Fish Crows are the more common type. On Cumberland Island, Pearson (1922: 88) reported Fish Crows as being common but did not record Common Crows, and on Sapelo Island, Teal (1959: 5) recorded the Common Crow as scarce in comparison with Fish Crows. In fact, until 1959 when young Common Crows were observed and collected on this island, the Fish Crow was believed to be the only nesting crow. Though the records are somewhat conflicting, apparently the same preponderance of Fish Crows exists in the Florida Keys (cf. Greene, 1946: 250; and Sprunt, 1954: 319, 321). As indicated above, it is

Fish Crow 63

apparent that the Fish Crow is less dependent upon man's agriculture for its food, but rather can fare well on native vegetable and animal material.

(Nevertheless, the Fish Crow *will* eat cultivated food products and is quick to learn of the readily available food at garbage dumps. Common Crows, furthermore, certainly eat much natural food. But, in a very general way, all of the available data show that Common Crows have increased in areas where man's agricultural activities abound, whereas Fish Crows seem to benefit less from such activities, unless it is a secondary result from the clearing of the land.)

Voice. The usual utterance of the Fish Crow is so distinctive that it is hardly to be confused with that of the Common Crow. Peterson (1958: 160) described its call as "a short, nasal *car* or *că*. Sometimes a two syllabled *că-hă*. . . . In late spring and summer, certain of the calls of young [Common] Crows are much like those of the Fish Crow." Sprunt and Chamberlain (1949: 373) wrote: ". . . it can always be recognized by its voice, which is very hoarse and raucous as though the bird were suffering with a bad cold." To Peterson's description above, I'd like to add that sometimes the birds give a nasal *cark*. Audiospectrograms for Fish and Common crows are presented by Davis (1958) for comparative purposes.

Morphology. Since this species occupies an extensive breeding range from Texas to at least Rhode Island, one might expect, *a priori,* geographic variations in measurements, or other bases for subspecific distinctions. With this possibility in mind, I first divided the adult males into two arbitrary north-south samples: (1) Virginia and northward, and (2) North Carolina and southward and the Gulf Coast region. The results of these computations are given in Table 2, and clearly show no significant clines or differences in the two geographic groups.

Since there were no significant differences between males in these two groups, further computations of measurements for this species were treated *in toto;* thus, in Table 3, measurements are from birds taken over the entire geographic range. By employing the formula for pooling data (MacArthur and Norris), the data for adult males as given in Table 2 are combined for their inclusion in Table 3.

Some description of the colors of the Fish Crow merit treatment here especially since a few authors have suggested taxonomic relationships between this North American form and the Mexican and Palm crows. First, in comparing *brachyrhynchos* with *ossifragus*, *ossifragus* has a much more glossy violet plumage and is more uniform in color. On *ossifragus* there are no flecks of violet-blue on the head, and very little of the scalation effect of the back which is so apparent in *brachyrhynchos*. (Sometimes a specimen of *ossifragus* will have a few scapulars or back feathers with this scalation effect.) The underparts are also more uniformly bluish-violet. Thus, the general appearance of the ventrum of *ossifragus* is uniform blue-violet, whereas the underparts of *brachyrhynchos* give a dull black effect.

There is little similarity in color between *ossifragus* and *imparatus* since the latter species is so richly glossed with purple or green. It is true that an occasional specimen of *ossifragus* might have a trace of the dorsal scalation effect, but this feature has not been detected in *imparatus*. Color comparisons between *ossifragus* and *palmarum* are more difficult; distinctions in not only color but other aspects of their comparative morphology are given under the discussion of C. *palmarum* to follow.

Interspecific relationships. As alluded to previously, one of the most interesting features of the Fish Crow is its geographic sympatry with the Common Crow over most of its range, and to students of evolution and avian ecology many important facts could be elucidated from an investigation of these sympatric relationships. Essentially, these relationships can be categorized as isolating mechanisms, those features of one species which prevent interbreeding with another species, for there has been no indication of hybridization between these two crows. These isolating mechanisms include ecologic, morphologic, reproductive, and other features (see Mayr, 1942: 247 ff., for a more complete classification of isolating mechanisms).

Morphologically, Fish and Common crows have been distinguished earlier in this paper on the basis of size and coloration (cf. Tables 1 and 3), the Common Crow being considerably larger and of a less glossy color than the Fish Crow. There is some overlap of the standard measurements when statistical treatments are employed. It is, of course, very probable that these morphological differences are sufficient to be *the* effective isolating mechanisms, but others are known. In the usual call notes, for example, mention has already been made of the significant audible difference between these two species. And ecologically, the two forms occupy somewhat different habitats, though recent investigations have demonstrated some interesting changes.

During the breeding season there are two aspects of microhabitat which deserve consideration. In the first place, there might be, locally, a tendency for Fish and Common crows to select different kinds of trees for their nests. In Maryland, for example, Fish Crows generally choose pines for nesting, but Common Crows show a preference for hardwoods. When one examines, however, a large series of nesting records from different parts of their ranges (for example, as given by Bent, 1946: 230 ff.), it becomes apparent that either species might utilize coniferous or deciduous trees, and it is probable that the selection of the nest site depends more upon factors other than just the kind of tree. Availability, height of tree, and the degree to which the nest may be concealed are no doubt important factors in selection of the nest site. An appropriate statement in this connection is that of H. H. Bailey (fide Bent, 1946: 261) who stated that Common Crows in Florida nest in "almost every kind of tree from mangrove, gumbo limbo, and cabbage palm of southern keys to pine, oak, and hardwood trees farther north." Nests of Fish Crows have also been reported in pines, hollies, cedars, oaks, and other deciduous types.

In the second place, there is a tendency for the Fish Crow to nest higher than the Common Crow. In the northeastern states, nests of the Common Crow have been reported from 10 to 70 feet, but those of the Fish Crow up to 90 feet. In South Carolina and Florida, Common Crows nest from 7 to 40 feet whereas Fish Crow nests have been found from 20 to 150 feet from the ground. The most complete records have been obtained from Maryland where the average height of 173 Common Crow nests was 36 feet, and the average height of 14 Fish Crow nests was 47 feet.

It is also of interest to note a general tendency for the Fish Crow to be more of a colonial nester, whereas this feature is less apparent in most populations of Common Crows (though well developed in northwestern, coastal populations). Bent (1946: 276) stated for the Fish Crow: "The nests are generally in small colonies, two or three pairs with their nests not far apart in a certain locality." And Burleigh (1958: 409) for this same species reported: "Where suitable nesting sites are available, two or more pairs will nest in close proximity to one another."

One last isolating mechanism between these two species—and perhaps the most effective one—is that of seasonal isolation. Mayr (1942: 251 ff.) cites several examples of seasonal isolation in birds, frogs, toads, and cuttlefish, emphasizing the effectiveness and importance of this isolating method. As Ressel (1889: 101), Burleigh (1958: 409), and other authors have pointed out, the Common Crow frequently nests as much as a month or more earlier than the Fish Crow. Figure 4 represents a compilation of several hundred nesting records for both species from the eastern and southern states, these records having been taken from numerous published works, especially state lists and Bent's compilations. The records were selected to represent, so far as possible, only nests containing fresh eggs, thus providing more comparable bases. In order to arrange these data comparably and to obtain a mean date of breeding for a given state, the days of the year were given consecutive, numerical values. Thus, January 1 = 1, January 2 = 2, February 1 = 32, March 1 = 60, May 12 = 132, and so on.

References from which nesting dates were obtained are as follows: Chapman (1932), Forbush (1925-29), Bent (1946), Sage, Bishop, and Bliss (1913), Eaton (1923), Cruickshank (1942), Bailey (1913), Murray (1952), Sprunt and Chamberlain (1949), Wayne (1910), Griffin (1940), Barkalow (1940), Burleigh (1958), Denton (1950), and Sprunt (1954).

Even though the data in Figure 4 show a good degree of reproductive isolation between these two species, all of these records might not be strictly comparable. Several authors have unequivocally stated that Common Crows are double-brooded (see Ressel, 1889; Bent, 1946; Sprunt and Chamberlain, 1949: 372), but there is no supporting evidence for this contention. A late nest does not *have* to be a second successful nest for the season, for, as Burleigh (1958: 407-8) correctly stated: "It is doubtful that more than one brood is reared each year, the few nests found in April and early May probably representing second attempts by pairs whose first nests were destroyed." A more valid criticism might lie in the fact that in some instances Common Crow nesting records were taken from the entire state (from mountains to seacoast with a wide range of climatic conditions) whereas Fish Crow records are generally from only the coastal strip where the climate would be less variable. And finally, the paucity of records from some states would tend to limit generalizations which might be induced from these records.

In spite of these possible objections, the facts indicate a strong degree of seasonal reproductive isolation between these two species, and certainly when this feature is coupled with those of morphology, voice, and habitat selection, it is easy to understand the absence of known hybridization between these two sympatric species.

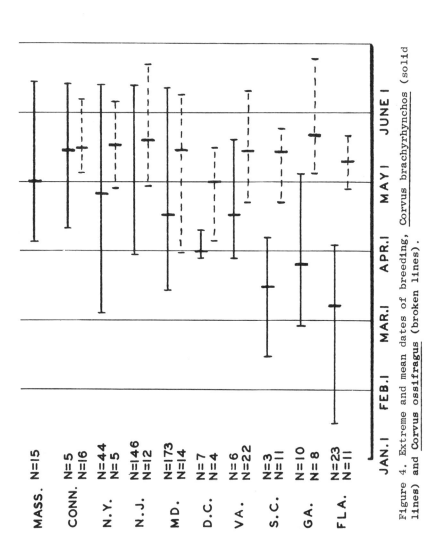

Figure 4. Extreme and mean dates of breeding, Corvus brachyrhynchos (solid lines) and Corvus ossifragus (broken lines).

TABLE 2

MEASUREMENTS OF ADULT MALE CORVUS OSSIFRAGUS FROM
NORTHERN AND SOUTHERN PORTIONS OF THE RANGE

WING

	N	Mean	S.E.	Range	S.D.	C.V.
Virginia, northward	16	287.8	1.22	268.0-298.5	4.89	1.7
North Carolina, southward	43	283.8	1.29	265.0-304.0	8.48	2.9

TAIL

| Virginia, northward | 16 | 157.6 | 1.77 | 142.0-166.0 | 7.07 | 4.5 |
| North Carolina, southward | 42 | 155.4 | 0.85 | 143.0-175.0 | 5.52 | 3.6 |

TARSUS

| Virginia, northward | 18 | 46.6 | 0.54 | 43.5-49.0 | 2.30 | 4.9 |
| North Carolina, southward | 46 | 46.5 | 0.15 | 40.6-50.0 | 0.99 | 2.1 |

BILL

| Virginia, northward | 16 | 30.8 | 0.62 | 27.0-34.2 | 2.49 | 8.1 |
| North Carolina, southward | 45 | 30.6 | 0.28 | 27.6-32.5 | 1.88 | 6.1 |

TABLE 3

MEASUREMENTS OF CORVUS OSSIFRAGUS

WING	N	Mean	S.E.	Range	S.D.	C.V.
males, adult	59	284.9	0.95	265.0-304.0	7.30	2.6
females, adult	47	270.7	0.99	260.0-293.5	6.79	2.5
males, first-year	6	267.6		260.0-275.0		
TAIL						
males, adult	58	156.0	0.82	142.0-175.0	6.24	4.0
females, adult	50	148.5	0.72	139.5-159.0	5.12	3.4
males, first-year	3	147.5		146.5-149.0		
TARSUS						
males, adult	64	46.5	0.35	40.6-50.0	2.83	6.1
females, adult	50	44.8	0.05	40.4-47.8	0.37	0.8
males, first-year	25	47.1	0.48	44.9-49.4	2.38	5.1
females, first-year	25	44.8	0.35	41.6-48.4	1.74	3.9
BILL						
males, adult	61	30.7	0.15	27.0-34.2	1.15	3.7
females, adult	50	28.3	0.23	25.7-31.1	1.63	5.8
males, first-year	24	29.9	0.33	26.3-31.7	1.61	5.4
females, first-year	24	27.4	0.39	25.0-31.6	1.89	6.9

5. MEXICAN CROW

Corvus imparatus Peters

Corvus imparatus Peters

Geographic Range
Mexico: specimens examined, 132: from Sinaloa, 18; San Luis Potosi, 8; Tamaulipas, 58; Nuevo Leon, 3; Sonora, 22; Nayarit, 22; and Veracruz, 1; others reported in the literature from Colima, Durango, and Maria Madre Island

Specifically, the range (Figure 1) along the eastern seaboard extends from the region of China, Nuevo Leon, and sixty miles south of Matamoros, Tamaulipas (Zimmerman, 1957), south to Panuco, Veracruz and Pujal, San Luis Potosi. On the western seaboard, this crow ranges from Agua Caliente on the upper Yaqui River, Sonora, inland to Tamazula, Durango, south to San Blas and Tepic, Nayarit, with additional records from Manzanillo, Colima, and Maria Madre Island.

Habitat. Many general and several specific statements are available in the literature on the habitat choice of this crow. In the state of Sonora, van Rossem (1945: 171) stated that this crow was "locally abundant, resident of the Tropical Zone . . . [and] tends to be concentrated in river valleys and farming districts, with avoidance of deserts and maritime associations." (The latter habitat avoidance is in conflict with Davis' statement below.) Sutton and Pettingill (1942: 23) reported Mexican Crows as common in the farming districts of Tamaulipas, and Sutton (1951: 102, 232) found these birds in Tamaulipas around cattle pens and in the lowlands. Blake's description (1953: 376) of the habitat was "in humid lowlands, especially river valleys." In another general account Miller (1957: 118) stated for this species: "Common resident of northern coastal provinces in Tropical

Zone, from sea level to at least 3000 feet, chiefly in farming districts." Davis gave more specific accounts of the preferred habitats by stating (1958: 158) that the crows of eastern Mexico occupy "semi-desert brushland (the towns, villages, farms, and ranch yards in the region are frequented as well as the brushy areas themselves); however, a few birds occasionally wander into open places in more humid woods at the limit of the usual habitat. Tall forests, true deserts, mountains, and the sea beach all are avoided. . . . Most of the birds are found between 100 and 1000 feet elevation but a few occur near sea level and up to about 1400 feet." (It is appropriate to mention here that numerous specimens from the coastal locality of Tampico [elevation 39 feet] have been examined in the present study.) For the population of crows of western Mexico, Davis (p. 161) reported them from "wet sand of the sea beach when the tide is out and along river estuaries; however, it also ranges back of the coast and up into the hills to elevations of 1000 feet or more. These birds are common in coastal towns and villages and also in the semi-desert deciduous woods some distance from the coast." Goldman (1951: 361, 378) recorded Mexican Crows from the Arid Upper Tropical Subzone and the Lower Austral (Sonoran) Zone, with specific occurrences being cited in both zones of the various states on east and west coasts.

From these varied descriptions it is perhaps dangerous to generalize on the habitat of the Mexican Crow, but a consensus would seem to suggest concentrations of the birds in farming districts not infrequently devoid of an aquatic association.

Voice. Descriptions of the voice of this crow in the literature are usually of birds representing the eastern population. Sutton (1951: 103), for example, described the Tamaulipan birds as having an "odd, surprisingly feeble cry which resembled the syllables *gar-lic,*" or "call-note a hollow, wholly unmusical *gar-lic* or *cow-rah*" (Sutton and Pettingill, 1942: 23). Davis (1958: 158) stated that the voice is "burry, low-pitched, and relatively low in volume, and sounds something like a frog croaking softly or someone plucking on a 'Jew's-harp.'" Crows from western Mexico are reported by Davis (1958: 161) to have a different voice from those in the east: "The call of the adult Sinaloa Crow is so startlingly different from that of the Tamaulipas Crow that a considerable difference in the morphology of the voice-making mechanism is at once suggested. The usual call is a clear 'ceow'." Also, an adult male calling in Nayarit was said to have a high, hollow, *caow* or *kwaa,* not nasal or guttural.

In an extensive audiospectrographic analysis of the voice differences between the eastern and western Mexican Crows, Davis (1958) has proposed that these allopatric populations should be designated as separate species. Whereas it is evident that these populations do have noticeably different calls, the question arises as to the interpretation of this fact relative to the biology of the populations and their systematic allocation.

A few general comments are appropriate in this connection. It must be remembered that the study of bird voices from the standpoint of individual and geographic variation and the interpretation of these variations is in an embryonic state, and, as far as most avian species are concerned, considerably more specific information needs to be amassed before broad conclusions should be drawn. Marler and Issac (1960: 124) suggest: "Not only are there differences between geographically separated populations of the same species, but there are also marked variations between individuals in the same population, and even in successive songs of the same individual. Although most attention has been directed toward geographic variation, it is difficult to describe such variation adequately until we have a clear picture of the song repertoire of the individual."

Notable among recent avian studies of this nature is that of Lanyon (1957) wherein he used audiospectrograms to assist in analyzing biological differences and similarities between eastern and western meadowlarks *(Sturnella)*. His study involved a careful consideration of different types of sounds (primary songs, rattles, chatters, and whistles). Further variations within one species were demonstrated later (Lanyon and Fish, 1958). In the final analysis, voice differences comprised only one of the isolation mechanisms between these sometimes sympatric species, because there are also differences in morphology, habitat choice, and the like. Geographic variations in songs of Traill's Flycatcher *(Empidonax traillii)* have been reported by Kellogg and Stein (1953), whereas Selander and Giller (1959) utilized voice differences of *Centurus* woodpeckers in their investigations of the interspecific biologic relationships of these birds. These studies represent beginnings in the use of modern techniques and the interpretations of bio-acoustics among birds.

In crows of any species, however, there is a paucity of precise scientific data available for inter- and intraspecific variations of calls and their meanings, so that the interpretation of the extant data must be used cautiously and in conjunction with other characteristics of the various species. Probably the most significant work toward the interpretation of crow voices has been that of Frings and his coworkers (Frings and Frings, 1959; Frings, *et al.*, 1958). One of their important contributions was the discovery of at least two types of calls in *C. brachyrhynchos*: (1) an assembly call and (2) an alarm call. In addition to these, Frings has numerous unanalyzed recordings of voices of nestlings, juveniles, and adults. In other words, Frings has attempted not only to discover inter- and intraspecific differences but has investigated the meanings of these differences by exposing various populations to the different calls.

No doubt isolating mechanisms of a visual or auditory nature in crows as well as other birds play important roles in species recognition, but I would agree with Sibley (1957: 172) that "before signal characters [releasers] can be fully utilized in taxonomy, it is necessary to determine their *function* and, for display movements and sounds, their *motivation*. . . ."

Morphology. When a series of adult Mexican Crow specimens is examined, some birds have greenish underparts but others are more purple. At first, it seemed that these color shades might be correlated with sex or geographic location, but closer scrutiny showed simply a seasonal correlation. Birds of both sexes collected between November and early March, regardless of location (Sonora, Nayarit, Tamaulipas), were distinctly purplish, but specimens from the same localities taken from April to June showed the greenish tendency. Evidently, this seasonal, gradual color change is related to chemical and/or physical change which is evident but less pronounced in some other species of crows (e. g., *C. palmarum*).

The Mexican Crow is the most highly colored of all the North American crows, since it is glossy black with a strong purplish appearance especially on the back. This distinctive lustrous coloration, not to mention its small size, affords immediate separation from any other species of crow. The back is completely devoid of the scalation effect which may be found in some other species.

In size *imparatus* is closest to *palmarum* but these two forms differ in color. Ventrally *imparatus* is solid metallic purple or purple-green whereas *palmarum* is a dull blue-black. On the back, *imparatus* is solid metallic purple, but *palmarum* is dull blue-black with a "suggestion of scalation."

In this connection, mention should be made of Hellmayr's allegation (1934: 5) that the Mexican Crow, "while easily distinguishable by smaller size and much more glossy plumage, is clearly conspecific with the North American Fish Crow." Thereafter, Hellmayr, Dorst and some others have referred to the Mexican Crow as *C. ossifragus imparatus*. Evidence amassed in the present investigation clearly shows that the exact opposite is the case, viz., the Mexican Crow and the Fish Crow are unrelated in virtually all of their features. There are significant differences in color, size, voice, and in habitat choice. The Fish Crow shows a decided preference for the vicinity of ocean beaches, river bottoms, and lake shores whereas the Mexican Crow is generally an upland bird reaching its abundance in cultivated farm regions. These striking ecologic and morphologic differences provide convincing evidence that the Mexican Crows are clearly not conspecific with either the Fish Crow or the Palm Crow.

Since Davis (1958) has proposed that the eastern and western populations of Mexican Crows be regarded as distinct species, it seemed desirable in this study to separate the mensural data for eastern and western samples. Accordingly, in Table 4 one can easily compare directly a given measurement for the two groups. It should be emphasized here that Davis' treatment of measurements was exceedingly scanty, was largely based upon Ridgway's data, and gave no distinctions between age groups. Thus, it is evident that the data in Table 4 are considerably more complete.

Table 4 shows that adult males from the eastern and western populations differ statistically only in tail length, a fact which Davis established by expressing wing/tail ratios. This type of ratio, incidentally, simply expressed more or less of a "constant" (wing length) in relation to a "variable" (tail length), and since tarsal and bill lengths are nearly identical for both populations, it would also be possible to compare these populations by using tarsal/tail or bill/tail ratios. (There is a tendency for the western birds to be larger in all of their measurements, even though, statistically, there is considerable overlap with eastern birds.) In adult females and first-year birds of both sexes, the measurements do not differ statistically. This would mean that the only mensural difference between eastern and western Mexican Crows is found in adult males and, in them, only in tail length.

If one recalculates wing/tail ratios from the more abundant data given in this paper rather than from Ridgway's meager data, somewhat different results are obtained than those presented by Davis (1958: 153, 161). In Table 5 ratios have been calculated only for adult males, and for representative samples (about 25) from the many available specimens.

These data, when plotted graphically (Figure 5) utilizing the "Dice graphs," reveal some interesting facts which, in a sense, corroborate Davis' findings. Adult male crows from western Mexico do have statistically significantly smaller wing/tail ratios than comparable forms from eastern Mexico, but in the various contiguous subspecific populations of *Corvus brachyrhynchos* one can find the same degree of differences. For example, crows from Ontario are significantly different from Florida and Alaska crows in wing/tail ratio. It would seem from this sort of comparison that the same difference (viz., a subspecific one) exists between eastern and western Mexican populations as does between Ontario and Florida populations. The only real difference lies in the fact that, on the one hand, we are dealing with allopatric Mexican birds, and, on the other hand, continuous populations of North American birds. Tail length per se (or wing/tail ratio) is not necessarily a good species' characteristic in just one group of crows, especially when similar variations in other populations of crows are considered carefully.

Since the present analysis is in conflict with Davis' taxonomic treatment of the Mexican Crows, some final settlement of the matter seems desirable. Actually, one is faced here with the basic question—What constitutes a bona fide species of crow? Davis contends that these Mexican crow populations differ in two principal respects—wing/tail ratio and voice—and that these features constitute species' differences. In the present work, however, it has been shown that these tail differences are present only in adult males, and that comparable differences are to be found among some subspecies of the Common Crow. Voice distinctions are somewhat more apparent, but I do not believe that this is sufficient evidence for delimiting a species, especially since the morphological differences are minute and there are essentially no ecological ones.

It might very well be true that the two Mexican crow populations represent incipient species, but from the present investigation of their characteristics—morphology, habitat and voice— it does not seem that they have yet reached the level of divergence attributed to bona fide species. This statement is made in spite of the fact that, as allopatric populations, geographical isolation is certainly evident, but, since they are so similar in their characteristics, it is purely conjectural at this point to suggest that voice differences, for example, would prevent interbreeding should they become sympatric or even contiguous populations.

Biologically and taxonomically it would seem wiser to represent the Mexican crows as subspecific populations, referring to the eastern Mexican Crow as *Corvus imparatus imparatus* Peters and the western Mexican Crow as *Corvus imparatus sinaloae* Davis. This treatment rather clearly indicates the true, close relationship between the two populations (morphologically and ecologically), whereas if they were regarded as distinct species, the slight differences of tail length in adult males and of voice would be magnified disproportionately in a biological sense. And a subspecific distinction is more in line with taxonomic treatments of other crow populations, agreeing generally with systematics of most other birds. For example Mayr (1948 fide Van Tyne and Berger, 1959: 358) stated, ". . . it is preferable for practical as well as for scientific reasons to treat all doubtful allopatric populations as subspecies. The scientific reason is that the mere fact that a population was unable to overlap the range of its nearest relative implies that it has been unable so far to develop isolating mechanisms that would permit coexistence."

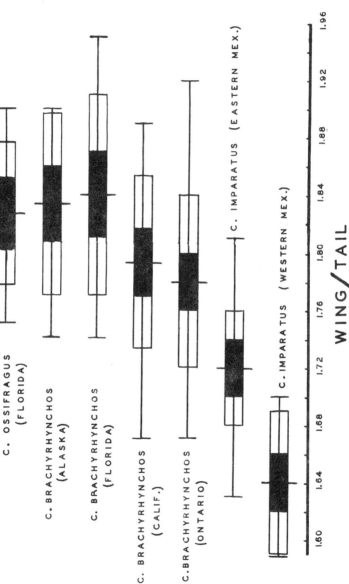

Figure 5. Graphic expressions of wing/tail ratios, selected crow populations. Data for adult males only. Horizontal lines are extremes; heavy vertical lines, the means; black rectangles, ± twice the standard error of the mean; open rectangles, ± one standard deviation.

TABLE 4

MEASUREMENTS OF CORVUS IMPARATUS

WING	Adult Males					
	N	Mean	S.E.	Range	S.D.	C.V.
Eastern population	27	250.5	1.44	231.0-265.5	7.46	3.0
Western population	24	253.5	1.50	242.0-267.0	7.35	2.9
TAIL						
Eastern population	26	145.9	1.46	132.0-155.0	7.45	5.1
Western population	25	155.0	1.12	147.0-164.0	5.60	3.6
TARSUS						
Eastern population	25	39.59	0.42	36.2-43.0	2.10	5.3
Western population	27	40.37	0.33	35.9-43.8	1.72	4.3
BILL						
Eastern population	25	29.11	0.35	25.2-32.4	1.74	6.0
Western population	25	29.97	0.24	27.4-32.5	1.19	4.0
WING	Adult Females					
Eastern population	10	238.4	1.32	232.5-246.0	4.17	1.8
Western population	23	238.3	1.52	229.5-262.5	7.26	3.1
TAIL						
Eastern population	10	140.4	0.96	136.0-147.5	3.04	2.2
Western population	22	144.0	0.99	133.5-150.0	4.62	3.2
TARSUS						
Eastern population	9	38.33		36.2-39.6		
Western population	23	38.68	0.30	34.8-41.6	1.44	3.7
BILL						
Eastern population	10	27.30	0.36	26.1-29.4	1.14	4.2
Western population	22	28.07	0.27	25.8-31.8	1.28	4.6
TARSUS	First-Year Males					
Eastern population	16	39.40	0.50	36.0-44.3	1.98	5.0
Western population	7	40.0		37.5-42.3		

TABLE 4 (continued)

MEASUREMENTS OF CORVUS IMPARATUS

BILL — First-Year Males

	N	Mean	S.E.	Range	S.D.	C.V.
Eastern population	17	28.83	0.26	26.6-30.4	1.09	3.8
Western population	6	28.50		26.5-30.9		

TARSUS — First-Year Females

	N	Mean	S.E.	Range	S.D.	C.V.
Eastern population	18	37.98	0.37	34.3-39.6	1.57	4.1
Western population	8	38.33		36.7-41.5		

BILL

	N	Mean	S.E.	Range	S.D.	C.V.
Eastern population	19	26.95	0.27	25.2-30.3	1.17	4.3
Western population	8	26.90		24.5-28.6		

TABLE 5

WING/TAIL RATIOS IN ADULT MALE CROWS

	N	Mean	S.E.	Range	S.D.	C.V.
Corvus ossifragus (Florida)	18	1.83	.01	1.75-1.90	0.05	2.7
Corvus brachyrhynchos (Ontario)	25	1.78	.01	1.67-1.92	0.06	3.4
Corvus brachyrhynchos (Florida)	25	1.84	.01	1.74-1.95	0.07	3.9
Corvus brachyrhynchos (Alaska)	24	1.83	.01	1.74-1.90	0.06	3.5
Corvus brachyrhynchos (California)	25	1.79	.01	1.67-1.89	0.06	3.4
Corvus imparatus (Eastern population)	26	1.72	.01	1.63-1.81	0.04	2.5
Corvus imparatus (Western population)	22	1.64	.01	1.59-1.70	0.05	3.1

6. CUBAN CROW

Corvus nasicus Temminck

Corvus nasicus Temminck

Geographic Range
Isle of Pines: specimens examined, 19: from La Vega, Pasadita, Caleta Grande, Jacksonville, Caleta Cocodrilos, and Santa Fe
Grand Caicos: specimens examined, 4
Bahamas: bone fragments reported by Bond (1956) from Crooked and Exuma Islands
Cuba: specimens examined, 62: from Guantánamo, Trinidad, Puerto Principe, Santa Rosa, Jimaguagu, Vertientes, Palo Alto, Panchita, Delirio, Santo Tomas, Sibanicú, Woodfred, Monte Verdi, Nicaro, Algarrobo, Ensenada de Cochina, and Bayate; others reported in the literature from Matanzas, Cárdenas, Camagüey, San Pablo, Cienaga, and western part of Piñar del Río

Habitat. This form, like some other species on various Caribbean islands, has been subjected to persecution by man, and now has a less extensive range (Figure 6) than formerly, though in some areas it is still locally abundant. Barbour's (1923: 106) description of its present habitat and status was as follows: "The Cuban Crow grows yearly less in numbers. Gundlach killed a few around Matanzas and Cárdenas as late as 1850, but for years they have been absent from these towns. The forests around Camagüey and the Trinidad hills had crows until 1915. . . . [They are] dependent on heavy forest, and disappear as it is cut."
Voice. The voice of *nasicus* has been described by Barbour as consisting of babbles and chatters in an infinite variety like the

Jamaican Crow. Bond (1956) suggests that it has voice different from the sympatric Palm Crow, having a variety of guttural, ravenlike sounds. And Vaurie (1957) recorded their notes as "croaking, guttural. . . . " These voice descriptions suggest some similarity to the White-necked Crow and to the more northerly ravens, but as indicated later in this paper, there are significant differences in size and habitat among these various forms.

Morphology. In color *C. nasicus* is closest to *C. brachyrhynchos* but specimens of the two can be separated since *nasicus* has less of the scale-like appearance on the back, and there is less of a contrast between the head-nape and the nape-back. The nape is about the same shade of deep violet gloss as the back of the head. In comparing these two species, the underparts of *nasicus* are usually more glossy. Other important morphological differences include the bare nostrils of *nasicus* (covered in *brachyrhynchos)* and the important size differences (see Table 6). Both sexes and ages of *brachyrhynchos* generally have longer wings, tails, and tarsi, whereas *nasicus* has a larger bill, all of these differences being statistically significant.

Specimens examined from the Isle of Pines could easily be separated into age groups on the basis of extreme wear and fading of rectrices and remiges in the first-year birds as well as the different shapes of the ends of the rectrices. Specimens from Cuba, however, were more difficult to separate into age groups because, for some unknown reason, there was less apparent wear and fading. Furthermore, the shape of the rectrices was less clear-cut as in other species of crows—"square-tipped" in adults vs. "pointed" in first-year birds. For this reason, the age of a few of the birds in the above analysis might have been judged incorrectly. Mayr (letter) has indicated similar difficulties in determining age of some of the Australian crows.

Too few specimens were available from the Isle of Pines and Grand Caicos to make valid comparisons with birds from Cuba. Therefore, in Table 6 specimens from all three of these islands are combined.

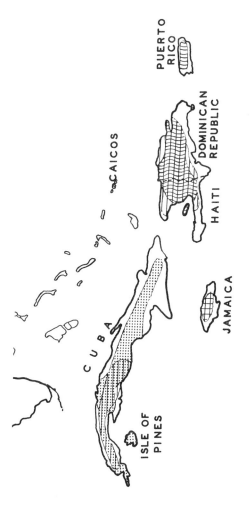

Figure 6. Approximate breeding distributions of four species of crows in the Caribbean Islands.

TABLE 6

MEASUREMENTS OF CORVUS NASICUS

WING	N	Mean	S.E.	Range	S.D.	C.V.
males, adult	34	274.2	1.54	257.5-293.0	8.99	3.3
females, adult	25	270.2	1.83	257.0-297.0	9.15	3.4
males, first-year	2			275.5-282.0		
TAIL						
males, adult	34	155.9	1.51	140.5-167.0	8.83	5.7
females, adult	24	152.1	1.50	138.0-161.0	7.37	4.8
TARSUS						
males, adult	36	51.2	0.55	45.2-56.7	3.32	6.5
females, adult	24	49.9	0.47	45.9-55.6	2.28	4.6
males, first-year	6	52.1		49.9-55.4		
females, first-year	11	49.8	0.46	44.2-52.2	1.54	3.1
BILL						
males, adult	36	42.2	0.32	37.9-46.9	1.91	4.5
females, adult	24	40.8	0.38	38.4-45.5	1.88	4.6
males, first-year	5	41.5		40.9-42.1		
females, first-year	11	40.6	0.55	38.1-43.4	1.83	4.5

7. WHITE-NECKED CROW

Corvus leucognaphalus Daudin

Corvus leucognaphalus Daudin

Geographic Range

Puerto Rico*: specimens examined, 11: from Sierra Luquillo, Santa Catalina (El Yunque); others reported in the literature from Lares, Quebradillas, Caguas, Utuado, and Río Mameyes

St. Croix*: repòrted from kitchen middens (Wetmore, 1927)

Hispaniola: 1. Haiti: specimens examined, 10: from Cerca la Source, Las Cahobes, Hinche, Dondon, and Cerca Cabrajal; others reported in the literature from St. Michel, Ennery, Port au Prince, Bois Laurence, Ganthier, Sources Puantes, La Selle, and Gonave Island. 2. Dominican Republic: specimens examined, 46: from Samaná and Samaná Bay, Almacen, Catare, Maiman, Magua, Sánchez, La Cañita, Sabaná de la Mar, Río San Juan, and Saona Island; others reported in the literature from Bonao, Riva, La Vega, Montecristi, Cibao Range, San Lorenzo, Port Rincón, Comendador, and Constanza

Habitat. According to the observations of Wetmore (1927) the White-necked Crow on Puerto Rico is a rare bird of the extensive, heavy forests, and disappears when these are cut. The localities mentioned above are principally inland (Figure 6) and

---

*Wetmore (1956: 93) has described a small crow *(C. pumilis)* from Puerto Rico and St. Croix. According to him, "it has not been found in living form, being known only from bones. Probably this small crow existed until modern times near Lares, Puerto Rico."

near mountainous, forested terrain, indicating this species' preference for that particular habitat. On Hispaniola, however, this crow is apparently more widespread since there are records of its occurrence from coastal mangrove swamps and cactus forests to the heavy, inland forests. The majority of these records from Hispaniola indicate that, here too, this crow shows a preference for the more inaccessible forests, and it is thought (Wetmore and Swales, 1931) that the clearing of the forests has contributed to a diminution in their numbers.

Voice. The White-necked Crow has a voice, according to Danforth (1929), which is "a most uncrowlike, gutteral sound." Wetmore (1927) described the Puerto Rican birds as having a high-pitched *klook* or a deep *wallough,* varied with guttural calls, and having no caws like "crows." On Hispaniola Wetmore and Swales (1931) reported "higher pitched notes . . . like those of the ravens of the north." Finally, Bond (1947) described their voices as being "raucous and ravenlike. A common utterance resembles a throaty *culik-calow-calow.*" Of all the crows studied in this report, the White-necked Crow seems to have the most uncrowlike series of notes, and the voice certainly suggests a relationship to the more northerly ravens (see discussion below).

Morphology. Ridgway (1904: 279) recognized a "Haitian Crow," *Corvus leucognaphalus erythrophthalmus,* stating that, compared to the Puerto Rican Crow *(C. l. leucognaphalus),* it was smaller (except bill), with larger feet, and had an iris varying from light reddish brown to bright orange red. Wetmore (1927) and Wetmore and Swales (1931: 327) made comparable measurements, and concluded that the size differences were negligible. Somewhat more data are given in Table 7, and even though the sample size from Puerto Rico is rather small, there do not seem to be any statistically significant differences between birds of the two insular populations. Interestingly enough, Hellmayr (1934) did not recognize these two subspecies, but, rather, further complicated matters by suggesting that the Cuban Crow *(nasicus)* was a subspecies of *C. leucognaphalus.*

There seems to be no evidence to support either of these contentions. In the first place, the Puerto Rican and Hispaniolan populations do not differ significantly enough in measurements to warrant their treatment as subspecies, and, in the second place, *nasicus* differs markedly from *leucognaphalus.* For example, *nasicus* always has exposed nostrils, whereas *leucognaphalus* many or may not have nostrils exposed. The underparts of *leucognaphalus* are more violet or purple than *nasicus*, and, of course, the former possesses many contour feathers with white bases. In size, *leucognaphalus* is significantly larger than *nasicus* (cf., Tables 6 and 7). The scale-like effect of the upper back is virtually absent in *leucognaphalus*, though it is, at best, poorly demonstrated on *nasicus*. It seems best, then, to retain the name *Corvus leucognaphalus* for all the White-necked Crows of Hispaniola and Puerto Rico.

Relationship to ravens. It is of singular interest to note the similarities between the White-necked Crows of these Caribbean islands and the White-necked Raven *(C. cryptoleucus)* of southwestern North America. It is tempting to suggest that the two might have been derived from a common ancestor, but lacking evidence for such a statement, it is perhaps better to think in terms of convergent evolution. In any event, these two forms are similar in that both have feathers with white bases on the lower throat, upper flanks, breast, nape (most conspicuous), back, and abdomen. No precise mensural studies were made of *C. cryptoleucus* here, but a superficial comparison with *leucognaphalus* showed that the former has a longer bill, heavier at the base, and a longer wing. *C. cryptoleucus* also is duller black and less glossy, and possesses pointed throat feathers, these being absent in *leucognaphalus*. *Cryptoleucus* has more numerous nasal bristles, pointing straight forward and always completely covering the nostrils. In *leucognaphalus* about one half of the specimens have the nostrils completely or partly exposed. Finally, the habitats of the two species are quite different, the White-necked Raven occupying arid deserts and the White-necked Crow being principally a forest species.

TABLE 7

MEASUREMENTS OF CORVUS LEUCOGNAPHALUS

WING

Adult Males

	N	Mean	S.E.	Range	S.D.	C.V.
Hispaniola	11	311.3	3.4	292.0-329.0	11.23	3.6
Puerto Rico	3	302.3		297.0-310.0		

TAIL

	N	Mean	S.E.	Range	S.D.	C.V.
Hispaniola	13	186.5	2.3	165.5-206.5	8.42	4.5
Puerto Rico	3	191.8		191.0-193.0		

TARSUS

	N	Mean	S.E.	Range	S.D.	C.V.
Hispaniola	12	54.1	0.15	50.2-57.2	0.52	0.9
Puerto Rico	3	50.1		48.3-51.0		

BILL

	N	Mean	S.E.	Range	S.D.	C.V.
Hispaniola	14	44.5	0.54	42.8-48.7	2.02	4.5
Puerto Rico	3	43.4		41.9-44.3		

WING

Adult Females

	N	Mean	S.E.	Range	S.D.	C.V.
Hispaniola	11	289.7	1.98	281.5-307.0	6.6	2.3
Puerto Rico	6	290.5		285.0-298.0		

TAIL

	N	Mean	S.E.	Range	S.D.	C.V.
Hispaniola	10	177.9	1.3	169.0-190.0	3.95	2.2
Puerto Rico	6	179.5		169.5-187.0		

TARSUS

	N	Mean	S.E.	Range	S.D.	C.V.
Hispaniola	12	49.0	0.71	45.4-52.4	2.44	4.9
Puerto Rico	6	49.6		47.3-52.6		

BILL

	N	Mean	S.E.	Range	S.D.	C.V.
Hispaniola	12	40.5	0.38	35.1-45.3	1.3	3.2
Puerto Rico	6	41.7		39.2-43.6		

WING

First-Year Birds, Hispaniola

	N	Mean	S.E.	Range	S.D.	C.V.
males	16	297.6	2.68	276.0-319.0	10.7	3.6
females	10	290.0	1.67	282.0-300.0	5.28	1.8

TABLE 7 (continued)

MEASUREMENTS OF CORVUS LEUCOGNAPHALUS

First-Year Birds, Hispaniola

TAIL	N	Mean	S.E.	Range	S.D.	C.V.
males	15	177.9	2.79	161.0-199.0	10.8	6.1
females	10	173.0	1.84	165.0-182.0	5.83	3.4
TARSUS						
males	16	53.1	0.69	49.0-56.2	2.77	5.2
females	11	52.0	0.85	49.9-56.0	2.81	5.4
BILL						
males	13	44.0	0.43	41.8-45.9	1.54	3.5
females	11	41.6	0.61	38.2-44.3	2.03	4.9

8. PALM CROW

Corvus palmarum Wurttemberg

Corvus palmarum Württemberg*

Geographic Range
  Cuba: specimens examined, 19: from Piñar del Río, Pinares, Panchita, Santa Rosa, Delirio, Guane, and Pedel River; others reported in the literature from Cienfuegos, Yaguaramas, Trinidad Hills, and Las Lomas de los Acostas
  Hispaniola: 1. Haiti: specimens examined, 19: from Port au Prince (40 mi. W.), Cerca La Source, l'Atalaye, Lake Assuei, Bombardopolis, Dondon, Ganthier, Hinche, Fond Parisian, and Trou Caiman; others reported in the literature from Mirebalais, Las Cahobes, Thomazeau, Gloré, Petit Goave, Cul de Sac Plain, Etang Saumatre, La Selle, Caracol, Artibonite, Gonaives, St. Michel, and St. Marc. 2. Dominican Republic: specimens examined, 51: from Constanza, Maniel, Tubano, Mt. Rusilla, Montecristi, Vasquez, Polo, and Espaillat Province; others reported in the literature from Cibao Range, Lake Enriquillo, Samaná Bay, and Las Matas

Habitat. On Cuba, Gundlach (about 1850) did not consider this crow very rare, but since his time it has been reduced (according to Barbour, 1923) to a few small bands in lonely, arid hills scattered with pines or around farms. More recently Vaurie (1957) studied some of these birds in the western part of Piñar del Río. The range is indicated on Figure 6.
  Bond (1947) reported that the Palm Crow was much more numerous on Hispaniola than on Cuba and that it was particularly

---

*Includes *Corvus minutus* Gundlach (see Ridgway, 1904: 276-7).

abundant in pine forests. Wetmore and Swales (1931) also recorded Palm Crows in "arid sections," "brushy regions," "rolling country," "about the tops of royal palms," and in swamps along Samaná Bay. Danforth (1929) recorded Palm Crows as being common in the rolling, hilly, sparsely vegetated country around Las Matas, in river sloughs, in tree cactus country, and in xerophytic forests. The distribution of this form and of the White-necked Crow on Hispaniola and of the Palm Crow and the Cuban Crow on Cuba indicate that these species-pairs are at least partly sympatric, though virtually nothing is known of interspecific relationships. In the pine forests of La Selle, *palmarum* is abundant, *leucognaphalus* apparently being absent. In the pine forests of northeastern Haiti, they are sympatric, with *palmarum* being the more abundant. Both occur near Port au Prince, *palmarum* being in the more open areas of the Cul de Sac Plain.

Voice. Notable differences of opinion exist as to the voice of the Palm Crow, for Danforth (1929) says "their caw was much like that of *Corvus brachyrhynchos,* but higher pitched, not so loud, and with a sort of burring, nasal quality." Wetmore and Swales (1931) compare the notes of the Palm Crow to those of the White-necked Crow by stating that the former has "a harsher caw, less musical than that of the white-necked species, resembling more the notes of the North American Crow." And Bond (1947) reports that the Palm Crow "utters a harsh *craa-craa* like North American Fish Crows or European Carrion Crows." Vaurie (1957) and Barbour (1923) both state that the voice resembles that of the Fish Crow. This latter comparison or suggested resemblance to *Corvus ossifragus* is of especial interest since, morphologically, there are some superficial resemblances between *ossifragus* and *palmarum.*

Morphology. Interestingly enough, Meinertzhagen (1926) considered the Cuban form *("minutus")* and the Hispaniolan bird *("palmarum")* as separate subspecies of *Corvus brachyrhynchos,* and Hellmayr (1934) and Meise (1928) made them subspecies of *Corvus ossifragus.* All of the evidence presented here clearly indicates that neither of these taxonomic treatments is correct. At this point it seems necessary only to mention the differences between the Common Crow *(C. brachyrhynchos)* and the Palm Crow *(C. palmarum):* there are conspicuous differences in size, color, voice, and habitat choice.

The differences between *C. palmarum* and the Fish Crow *(C. ossifragus)* require elucidation here, because, in general, these two forms are superficially similar. The question remains then—should *palmarum* be considered as a subspecies of *ossifragus* as Hellmayr avers? It is my contention that it should not be, because there are several notable differences between the two. Insofar as size is concerned *ossifragus* has a much longer wing and tail, and *palmarum* has a longer tarsus and bill, the latter also being heavier at the base. In color, *ossifragus* is more blue-black (i. e., more iridescent) ventrally than *palmarum*, the latter bird tending to be blacker. Dorsally, it is somewhat more difficult to distinguish the two birds, but, again, *ossifragus* tends to be more blue-black, and *palmarum* blacker. In voice, some authors, as suggested above, indicate similarities between *palmarum* and *ossifragus* whereas others indicate a resemblance between *palmarum* and *brachyrhynchos* or *leucognaphalus*. This is, obviously, a feature which requires further study. With regard to habitat, there is no question that *ossifragus* is principally a bird of shore lines, lakesides, and river drainage systems, whereas *palmarum* is *principally* an upland form (see habitat descriptions above).

In summary, it is evident that the Palm Crows from Cuba and Hispaniola represent a species distinct from both *ossifragus* and *brachyrhynchos*.

Subspecies. Mensurally, the most complete data for comparing the two insular populations are available for adult males, and by examination of these figures (Table 8) it can be seen that the birds from Cuba tend to have larger tarsi and smaller bills. By comparing these figures statistically, i. e., utilizing standard errors, it is apparent that adult males from Cuba and Hispaniola differ only in tarsal and bill lengths. Since adult females (from Cuba) and first-year birds of both sexes are poorly represented in these samples, differences among these ages and sex are not clear. On the basis of measurements alone, using both sexes and both age groups, there are 16 possible points for statistical differences. Discounting the first-year birds, in these analyses there are only two out of eight statistical differences, viz., tarsal and bill lengths of adult males. It does not seem desirable or scientifically accurate to name these as two distinct subspecific populations on the basis of these few mensural differences.

Wetmore and Swales (1931: 330) stated that "the small crows found on Hispaniola and Cuba are similar in size and seem to differ only in color. . . . " In fact, on the basis of a supposed color difference, the Palm Crow of Cuba was given the name

Corvus palmarum minutus Gundlach, and that of Hispaniola the name *C. p. palmarum* Württemberg. Wetmore and Swales, Ridgway, Bond, Bangs and Peters, Meinertzhagen, Dorst, and other authors have suggested that the Hispaniolan and Cuban birds differ in color, and have given different names to the two populations on this basis. In addition to the supposed color difference, Dorst (1947: 73) stated that the Hispaniolan birds have a longer and narrower bill. According to the present analysis, he was correct, at least insofar as measurements are concerned in adult males.

According to Wetmore and Swales (1931) and Bond (1947: 156) the Cuban birds are supposedly blacker, or the Hispaniolan birds have more violet iridescence. These authors evidently had extremely small series of specimens which were undoubtedly not separated into correct age groups, sexes, and most important of all, seasonal groups. Five adult males from Cuba were compared by me with 7 from Hispaniola. The Cuban birds were taken as follows: 1 in December, 1 in February, and 3 in April. Hispaniolan birds were taken: 2 in September, 1 in February, 3 in March, and 1 in April.

By considering just the Hispaniolan birds first, dorsal views revealed the fact that the birds taken in the fall were strikingly iridescent (purple and violet showing especially on the tail, upper back, secondaries, and wing coverts), but passing down the series to the April bird, two things were evident. First, the iridescence was markedly reduced, though not completely absent, and the result was a blacker (April) individual. Second, there was wear and apparent destruction of pigment especially in the rectrices and remiges as these feathers became noticeably browner.

Examining, then, the series of 5 males from Cuba in the same seasonal progression, I noted that the December specimen was more iridescent than those taken in February and April. The latter were blackish, but still showed a trace of the iridescence in the remiges. For direct comparison, the December male from Cuba was placed among the September birds from Hispaniola, and none of these could be distinguished from one another. Furthermore, birds from both islands taken in the spring were intermixed, and, once again, it was impossible to separate birds on the basis of color.

Precisely the same results were obtained by examining the adult females, although in this instance the sample size was even smaller. The one adult female Cuban specimen taken in

December was highly iridescent and had a fresh plumage; it was comparable in color with birds of the same sex and age taken on Hispaniola in September but was more iridescent than specimens taken in the spring from either island.

Due to more pronounced wear and fading, first-year birds were not considered in this particular comparison.

The immediate result of this colorimetric comparison seems to be that one needs to exercise extreme caution in comparing Palm Crows, taking special care to select birds from comparable seasons. In both insular populations, fall or early winter specimens are not worn and are iridescent, but as the season progresses, physical and/or chemical changes affect pigmentation so that the feathers appear browner at their tips, are generally worn, and give the over-all appearance of a blacker bird. Consequently, if one compares a fall specimen with a spring specimen within or between either of these two insular populations, the one bird would be more iridescent and the other blacker.

It seems conclusive, then, that the Cuban Palm Crows and the Hispaniolan Palm Crows do not differ in color. Furthermore, there are too few mensural differences to warrant subspecific distinctions. As a result of these studies, I would suggest that the Palm Crows be "returned" to their monotypic status, *Corvus palmarum* Württemberg.

TABLE 8

MEASUREMENTS OF CORVUS PALMARUM

	N	Mean	S.E.	Range	S.D.	C.V.
WING		Adult Males				
Hispaniola	19	251.4	1.72	243.0-262.0	7.49	3.0
Cuba	10	253.1	1.68	245.0-264.0	5.30	2.1
TAIL						
Hispaniola	18	141.2	0.91	134.5-152.0	3.85	2.7
Cuba	10	141.9	1.18	135.0-148.0	3.73	2.6
TARSUS						
Hispaniola	19	49.5	0.30	46.5-52.8	1.32	2.7
Cuba	11	53.9	0.99	49.0-61.1	3.27	6.1
BILL						
Hispaniola	20	35.1	0.28	32.8-37.7	1.23	3.5
Cuba	11	32.3	0.38	30.5-34.2	1.25	3.9
WING		Adult Females				
Hispaniola	23	247.0	1.65	231.0-263.0	7.89	3.2
Cuba	2			237.0-245.5		
TAIL						
Hispaniola	23	140.6	0.96	131.0-151.0	4.60	3.3
Cuba	2			134.5-140.0		
TARSUS						
Hispaniola	24	48.5	0.40	45.9-53.2	1.94	4.0
Cuba	2			49.5-50.2		
BILL						
Hispaniola	23	33.2	0.42	30.9-38.5	1.99	6.0
Cuba	2			29.5-31.3		
TARSUS		First-Year Males				
Hispaniola	8	49.1		45.0-51.6		
Cuba	1	49.7				

TABLE 8 (continued)

MEASUREMENTS OF CORVUS PALMARUM

BILL

First-Year Males

	N	Mean	S.E.	Range	S.D.	C.V.
Hispaniola	9	33.1		28.7-36.2		
Cuba	1	31.9				

TARSUS

First-Year Females

	N	Mean	S.E.	Range	S.D.	C.V.
Hispaniola	7	48.6		45.7-52.9		
Cuba	3	53.6		52.6-54.3		

BILL

	N	Mean	S.E.	Range	S.D.	C.V.
Hispaniola	7	33.4		31.9-35.6		
Cuba	3	32.5		32.3-32.6		

9. JAMAICAN CROW

Corvus jamaicensis Gmelin

Corvus jamaicensis Gmelin

Geographic Range
　Jamaica: specimens examined, 20: from Windsor, Priestman's River, Moneague, and Spanishtown; others reported in the literature from Lumsden

Habitat. Little has been published on the habitat of this species, but Gosse (1847) reported that it was found only in the wildest parts of the mountainous regions and never below 2000 feet elevation. Although in Gosse's time the Jamaican Crow might have occurred only at high elevations, this statement is not completely accurate, for specimens have been taken at lower elevations since his time (see above localities and Figure 6). Danforth (1928) and Bond (1956) generally concurred that this crow is to be found in the wild, wooded hills, but is also common in fairly open, settled country around Moneague. Taylor (1955: 13) reports it as being restricted to St. Ann, Trelawny, and the Cockpit country.

Voice. Locally the Jamaican Crow is also known as the Jabbering Crow or Chattering Crow, these names suggesting the character of its voice. Gosse described it as being garrulous, and Bond said that "its voice resembles a harsh *craa-craa.*" It also has a raven-like croaking, suggesting the voice of *nasicus*, and other notes resemble those of *palmarum*. Its note resembles the caw of European rooks according to Taylor.

Morphology. In measuring specimens of the Jamaican Crow, I utilized birds taken in any month provided there was no evidence of wear. This procedure was followed because there is hardly any possibility that these birds were not from the area of breeding. Furthermore, as was true in some of the other Caribbean crows, it was not always possible to distinguish clearly between the age groups, and for this reason, an insignificant number of the birds listed as adults might, in fact, have been first-year birds.

This is the only form of Caribbean or North American crow over which there has been little disagreement on its nomenclature. In previous listings or revisions by Meinertzhagen, Dorst, Hellmayr, and others, the Jamaican Crow has been universally referred to as *Corvus jamaicensis,* and this taxonomic treatment, I feel, has been due to the unique, almost uncrowlike appearance of this species. The Jamaican Crow is the dullest of all Caribbean crows, being sooty black and having little gloss or iridescence at any season. There is no scalation to the back. The wings, tail, and neck are slightly violet, whereas the underparts are dull sooty black, a little browner than above. About one-half the specimens examined had nostrils fully exposed, whereas the nostrils of others were either partially exposed or fully covered. Many specimens showed a bare area underneath and behind the eye and at the angle of the bill, but Dorst (1947) in his key to the species of *Corvus* maintained that the post-ocular region of *jamaicensis* was feathered.

In size, *jamaicensis* is the smallest of the Caribbean crows (Table 9), and, with the exception of *imparatus,* is the smallest of all crows found in the Nearctic Realm.

TABLE 9

MEASUREMENTS OF CORVUS JAMAICENSIS

	N	Mean	Range
WING			
males, adult	8	235.4	231.0-239.5
females, adult	6	231.6	220.5-246.0
males, first-year	3	223.3	218.0-223.5
TAIL			
males, adult	8	141.9	135.0-148.5
females, adult	5	140.8	133.0-149.0
males, first-year	2	129.2	125.5-133.0
TARSUS			
males, adult	8	46.8	44.5-51.9
females, adult	5	45.4	44.7-45.8
males, first-year	4	45.1	42.2-48.7
females, first-year	2	43.6	42.5-44.7
BILL			
males, adult	7	34.1	32.4-35.3
females, adult	6	32.7	31.1-34.9
males, first-year	4	33.3	31.8-36.5
females, first-year	2	33.5	32.9-34.1

10. OTHER INTERSPECIFIC RELATIONSHIPS

Weights

The use of body weights for suggesting taxonomic relationships in birds must be handled cautiously due to the many variables, not the least of these being fat content (Johnston and Haines, 1957). With a bird as large as a crow, it was not feasible to obtain a fat-free weight by extracting fat chemically, but since all specimens used were spring or summer birds, it is probable that fat deposits were at a minimum. Weight values presented in Table 10 for *C. brachyrhynchos* (Georgia and Washington) and *C. ossifragus* were obtained personally from fresh specimens in the field. The remaining weights were transcribed from specimen labels.

The data show in every instance that adult males tend to outweigh adult females; in *brachyrhynchos* the difference is about 50 grams. Insufficient weights were available for first-year birds, but for Georgia the following data were obtained: 20 males (mean=421.3; range=368-466) and 17 females (mean= 387.1; range=324-438.3). It is evident from comparisons here that the first-year birds are lighter in weight than adults of comparable sex, a statement which closely correlates with the results obtained from linear measurements.

Whether the data in Table 10 can be used to suggest intraspecific relationships is debatable, due especially to the variables in sample size and techniques. Some trends are evident, however, for the eastern populations of Common Crows in the South are lighter in weight than birds from farther north. A similar trend is not evident in western populations, except when the California and interior British Columbia samples are compared.

TABLE 10

WEIGHTS OF ADULT CORVUS BRACHYRHYNCHOS AND CORVUS OSSIFRAGUS

	Adult Males		Adult Females	
	N	Mean and Range	N	Mean and Range
C. brachyrhynchos				
Georgia	33	446.7(396-513)	37	396.6(359-444.3)
Ontario and Michigan	11	504.4(434-575)	9	427.7(390-486)
California	17	417.8(346.5-460.2)	11	368.5(316-420)
Western Wash.	19	413.3(326-486.3)	7	364.3(314.6-421.2)
Alaska	10	414.2(388.2-444.1)	4	347.8(345.4-350.7)
Interior B. C. and Idaho	8	449.3(432.9-475.2)	5	415.4(383.6-461.6)
C. ossifragus				
Georgia, Florida and South Carolina	5	295.3(255-326)	7	282.6(268-294)

TABLE 11

OCCURRENCE OF MALLOPHAGA ON AMERICAN CROWS

	*Philopterus oscellatus osborni	*Brüelia rotundata	†Brüelia perwienae	‡Myrsidea americana	‡Philopterus corvi
C. b. brachyrhynchos	X‡			X	X
C. b. pascuus	X	X			
C. b. hesperis	X	X			
C. b. caurinus	X‡	X			
C. ossifragus	X	X		X	X
C. nasicus	X		X		
C. imparatus	X				
C. palmarum		X			

* Identifications by K. C. Emerson.
† Reported by Ansari (Bull. Br. Mus., Ent., 5: 145-82. 1957)
‡ Also reported by Peters (1936: 20)

Host-Parasite Relationships

In general American crows handled in this study were not heavily infested with ectoparasites, and those identified here were obtained from specimens in the American Museum of Natural History. Mallophaga were the only ectoparasites which could be shaken dead from the specimens. Subsequently, about 87 of these chewing lice were mounted and sent to K. C. Emerson who kindly provided most of the identifications in Table 11.

The data, though somewhat inadequate for the Caribbean crows, presented in Table 11 do not support the thesis, held by some, that ectoparasites are host specific and have evolved with their hosts. Only in *C. nasicus* has a distinctive, host-specific louse been described, and it is possible that a more thorough examination of *C. palmarum* might reveal *Brüelia perivienae* on that crow too. The fact that *Brüelia* is a large genus and in need of revision like *Myrsidea*, would tend to cast additional doubt on the use of these ectoparasites in suggesting host relationships among American crows. Host-parasite relationships have been useful, however, in suggesting relationships of some other birds (nuthatches—Norris, 1958: 261; European crows—Emerson, letter).

Phylogeny

From the foregoing species' analyses, it has been demonstrated that the American crows comprise a group of seven species, each of which is definable on morphologic and ecologic traits. These seven species generally fall into two size groups: (1) large: *C. brachyrhynchos, leucognaphalus,* and *nasicus;* (2) small: *ossifragus, palmarum, imparatus,* and *jamaicensis,* with *ossifragus* being the largest of the small forms. Whether these two groups comprise natural biological units, the species showing affinities one for another, is debatable, but in some of their morphologic features, *brachyrhynchos, nasicus,* and *leucognaphalus* are all similar. The small crows also exhibit some general similarities; especially is this true between *ossifragus* and *palmarum,* though *jamaicensis* in its dull dress differs notably from the glossy *imparatus*. Granted that these two size groups are not entirely artificial, we may then hypothesize on the phylogeny and spread of these American forms.

Bond was at least partly correct when he stated (1934: 343) that "the West Indies are Nearctic in their bird life" especially

Other Interspecific Relationships

as far as *Corvus* is concerned. But Mayr (1946: 19) regarded "the crow family . . . [as] examples of Old World groups which have established minor secondary evolutionary centers in North America. . . ." If these two ideas are correct, *Corvus* evidently reached North America via a Tertiary land bridge across the Bering Strait. In this regard mention should be made of a statement by Fay and Cade (1959: 128): "In the recent history of St. Lawrence Island there is a verbal record of a small *Corvus*, a crow of some kind." Even today there are similarities between some Asiatic and American crows *(C. corone* and *brachyrhynchos)* and among the ravens *(C. corax).*

It is likely that an ancestral, adaptable crow evolved in North America into the currently proposed large and small types. The large type spread into the islands of the Caribbean area, and established separate insular populations. These populations then became reproductively isolated through the centuries, and gradually became differentiated into the contemporary types, *nasicus* and *leucognaphalus*. These latter insular forms developed into forest types as contrasted with their North American relative *(brachyrhyncios)* which has become in historical times more adaptable to areas cleared for man's agriculture.

From the same ancestral type there evolved a smaller type of crow which spread into Mexico, via the coasts, and into the Caribbean areas. In the long period of reproductive isolation on the Caribbean Islands, these insular crows became differentiated into the forest types of today, *palmarum* and *jamaicensis*. The Mexican Crow has developed somewhat away from the coastal environments, but the Fish Crow has retained this coastal affinity more than any of its near relatives.

As far as the time element is concerned, we might speculate further by suggesting that those species of crows which are most distant from the North American continent today are the most differentiated types from the ancestor, and vice versa, and that these have experienced the longest period of isolation. *C. jamaicensis* and *imparatus* are the two most striking small crows, and except for one population of *palmarum* are the farthest from *ossifragus* on the North American continent. Conversely, *palmarum* (of Cuba) is closest to *ossifragus,* spatially and morphologically. A similar pattern is seen in the large species, for *brachyrhynchos* most closely resembles the Cuban Crow *(nasicus)* but is more different in appearance, habits, and other features from *leucognaphalus*.

The contemporary, sympatric relationships demonstrated in

this paper are notable at this point, too, for on all the large Caribbean islands (except Jamaica) where crows exist today, there are (or were in the recent past) both a large and a small species of crow (recall *C. pumilis* of Puerto Rico). As implied above, it hardly seems possible that both species on the same island originated from the same immediate ancestor, but, rather it seems more plausible to suggest a polyphyletic origin of these Caribbean types. On the North American continent, the sympatric *brachyrhynchos* and *ossifragus* are evidently species resulting from a secondary contact. It is possible that the ancestral crow gave rise to a small type which inhabited coastal environments, and as the coastline was extended a hundred or more miles (the present Coastal Plain), this small, coast-loving species became isolated from the upland, larger type.

Finally, the present studies shed some light on the question as to why there are no crows in South America. If *Corvus* reached North America via an Asian land bridge, then obviously it would spread southward through Mexico and Central America to northern South America. Now, the most plastic of the American crows, *C. brachyrhynchos,* is not a forest type. If this feature of its habitat choice were a limiting factor, it could be easily explained why it could not penetrate and inhabit the dense, unbroken forests of Central and South America.

One might argue, of course, that *imparatus* has evidently moved southward along a coastal route and that many of the Caribbean crows *are* forest types. *C. imparatus,* however, exists only in the semi-arid uplands, not in dense forests. Given a considerable length of time, it might spread southward along the coast but would seemingly be closely restricted to the coasts where dense forests are contiguous. Then, as in the case of the forest-inhabiting crows of the Caribbean islands, we might simply suggest that the ancestral crows extended just that far southward before becoming geographically and reproductively isolated. Being isolated, it was a matter of occupying the forest niche or becoming extinct. It would seem, then, that the choice of habitat, perhaps coupled with the time element, has been the most distinctive limiting factor toward the southward spread of crows into Central or South America.

11. SUMMARY AND CONCLUSIONS

The present investigation deals with a systematic and biologic analysis of the crows inhabiting North America and some of the Caribbean islands. Morphologically, these seven species of *Corvus* differ from one another in linear measurements (wing, tail, tarsus, and bill from nostril) and/or color, these features being examined on 2269 breeding specimens. Biologically, differences in habitat choice, voice, reproductive phenology, and migration have been analyzed. Of paramount importance in the study has been the constant recognition of both sex and age groups within a given species.

The Common Crow *(C. brachyrhynchos)* of North America is the most widespread and morphologically variable of all the forms studied here. A statistical examination of its standard measurements led to the conclusions that (1) *C. b. paulus* is not sufficiently different from *C. b. brachyrhynchos* to warrant its recognition as a distinct subspecies, (2) the proposed *C. b. hargravei* of Phillips is indistinct from contiguous populations of the Common Crow, (3) the breeding range of *C. b. brachyrhynchos* should extend westward to Arizona, Colorado, Montana, and Alberta, and (4) the range of *C. b. hesperis* should include California, Oregon, eastern Washington, Idaho, and eastern British Columbia. Also, an intensive field and specimen study of the Northwestern Crow was undertaken to elucidate both morphologic and ecologic traits of this form, and it was concluded that it is a well-marked subspecies of the Common Crow, being properly identified as *C. b. caurinus*. Evidence for this conclusion was forthcoming upon the discovery of a broad zone of intergradation in southwestern Washington where specimens intermediate in measurements and voice were noted. Thus, the range of the Northwestern subspecies extends from Alaska southward along coastal British Columbia to northwestern Washington; in mid- and southwestern Washington, and at least in the Fraser River valley of British Columbia, it intergrades with the Western subspecies, *C. b. hesperis.*

The Fish Crow *(C. ossifragus),* principally a littorine and riparian form of the southern and eastern United States, differs from *brachyrhynchos* in size, coloration, voice, usual habitat choice, and reproductive phenology. It is particularly this latter feature which is believed to be among the most effective isolating mechanisms between these two species. Evidence is amassed to show that *ossifragus* has spread in recent years northward along the Atlantic Coast, inland out of the Coast Plain into the Piedmont Plateau especially along river systems, and into more xeric habitats where conditions are favorable.

C. imparatus, the Mexican Crow, is specifically distinct from *ossifragus* on the basis of differences in size, color, voice, and the preferred habitat. Analysis of characteristics of the allopatric populations of *imparatus* revealed the fact that the differences (size and voice) between them are of a subspecific nature, so that the eastern population should be referred to as *C. i. imparatus* and the western one as *C. i. sinaloae.* Interspecific color variations are of a seasonal nature with winter specimens being purplish and spring specimens being greenish ventrally.

The Cuban Crow *(C. nasicus)* most closely resembles *brachyrhynchos* in its color and size, but may be distinguished from the Common Crow by careful analysis of color shades, voice, and habitat, since *nasicus* is primarily an inhabitant of heavy, undisturbed forests. Furthermore, statistical analyses of the measurements of these two species showed some significant differences.

Another large crow, *C. leucognaphalus* (White-necked Crow) occupies heavy forests of Puerto Rico and Hispaniola. In its voice and feathers with white bases, it resembles the North American White-necked Raven *(C. cryptoleucus)* but differs from this form in size, coloration, habitat choice, and perhaps other features. The two insular populations (Puerto Rico and Hispaniola) are morphologically indistinguishable, thus leading to the contention that *leucognaphalus* is a monotypic species.

C. palmarum, the Palm Crow, occurs on Cuba and Hispaniola where it occupies a variety of habitats. Its morphological characteristics suggest a relationship to *ossifragus* and/or *imparatus,* but by measurements and/or color these three species can

Summary and Conclusions

be easily distinguished. Palm Crows of Cuba and Hispaniola are shown to be indistinguishable on the basis of color or measurements. Seasonal color differences are demonstrated: fall and early winter specimens are more iridescent whereas those taken in the late winter and spring are blacker dorsally. Thus, *palmarum* should also be a montypic species.

The dullest of all the American crows is the Jamaican species, *C. jamaicensis*, and with the exception of *imparatus* is the smallest. Its unique voice and morphology warrant its recognition as a distinct species.

Weights of 171 adult *brachyrhynchos* from different geographic areas were analyzed for possible differences, but the only consistent differences were associated with sex and age. Adult males tend to be heavier than adult females and first-year birds are lighter than adults of the same sex.

By utilizing the available Mallophaga from the various species of crows, host-parasite relationships were investigated. In only one instance was a host-specific louse discovered, with the remainder of the lice species being randomly distributed on the various crows.

Based principally on size and color, the seven species generally fall into two groups: (1) *brachyrhynchos - nasicus - leucognaphalus* and (2) *ossifragus - palmarum - jamaicensis - imparatus*. It is suggested that a monophyletic ancestor gave rise to these contemporary species, some of which have become reproductively isolated and morphologically distinct insular forms. Finally, it is proposed that crows have not extended southward into Central and South America due to the barriers of tropical and subtropical forests, especially since the nearest crow *(imparatus)* is not typically a forest inhabitant.

LITERATURE CITED

Allen, Glover M. 1903. A list of the birds of New Hampshire. Proc. Manchester Inst. Arts and Sci., 4:23-222.
Amadon, Dean. 1944. The genera of *Corvidae* and their relationships. Amer. Mus. Nov., 1251:1-21.
–––––. 1949. The seventy-five percent rule for subspecies. Condor, 51:250-58.
–––––. 1950. The species–then and now. Auk, 67:492-97.
American Ornithologists' Union Committee. 1957. Check-List of North American Birds. Fifth Edition. 691 pp.
Bailey, H. H. 1913. The Birds of Virginia. J. P. Bell Co., Lynchburg, Va. Pp. 193-97.
–––––. 1923. The status of the Florida Crow *(Corvus brachyrhynchos pascuus)*. Wils. Bull., 35:148-49.
Bailey, Florence M. 1928. Birds of New Mexico. N. M. Dept. Game and Fish. 807 pp.
Baird, S. F., T. M. Brewer, and R. Ridgway. 1875. A History of North American Birds, Vol. II. Little, Brown and Co., Boston. 596 pp.
Barbour, Thomas. 1923. The birds of Cuba. Mem. Nuttall Ornith. Club, 4:106-7.
Barkalow, Fred S. 1940. Additional notes on nesting extremes for birds breeding in the Atlanta region. Oriole, 5:53-54.
Bent, A. C. 1946. Life histories of North American jays, crows and titmice. U. S. Nat. Mus. Bull., 191:1-495.
Bishop, L. B. 1900. Birds of the Yukon region. N. Amer. Fauna, 19:47-100.
Blair, W. F., A. P. Blair, P. Brodkorb, F. R. Cagle, and G. A. Moore. 1957. Vertebrates of the United States. McGraw-Hill Book Co., Inc., New York. 819 pp.
Blake, E. R. 1953. Birds of Mexico. University of Chicago Press, Chicago. 674 pp.

Bond, James. 1934. The distribution and origin of West Indian avifauna. Proc. Amer. Phil. Soc., 73:341-49.
─────. 1947. Field Guide to Birds of the West Indies. Macmillan Co., New York. 257 pp.
─────. 1956. Check-List of Birds of the West Indies. Acad. Nat. Sci. Phil. 4th Edition. 184 pp.
Bowles, J. H. 1900. The northwest crow. Condor, 2:84-85.
Brooks, Allan. 1942. The status of the northwestern crow. Condor, 44:166-67.
Brooks, A., and H. S. Swarth. 1925. A distributional list of the birds of British Columbia. Pac. Coast Avif., 17:80-81.
Burleigh, Thomas D. 1941. Bird life on Mt. Mitchell. Auk, 58:334-45.
─────. 1958. Georgia Birds. University of Oklahoma Press, Norman, Okla. 746 pp.
Chapman, Frank M. 1932. Handbook of Birds of Eastern North America. D. Appleton and Co., New York. 581 pp.
Connell, Clyde E., E. P. Odum, and Herbert Kale. 1960. Fat-free weights of birds. Auk, 77:1-9.
Cruickshank, A. D. 1942. Birds around New York City. Amer. Mus. Nat. Hist. Handbook Ser. No. 13. 489 pp.
Danforth, S. T. 1928. Birds observed in Jamaica during the summer of 1926. Auk, 45.
─────. 1929. Notes on the birds of Hispaniola. Auk, 46:371.
Davis, L. Irby. 1958. Acoustic evidence of relationship in North American crows. Wils. Bull., 70:151-67.
Denton, J. Fred. 1950. The Fish Crow breeding in McDuffie County, Georgia, a further extension of its range. Oriole, 15:33.
Dice, L. R. and H. J. Leraas. 1936. A graphic method for comparing several sets of measurements. Contr. Lab. Vert. Gen., 3:1-3.
Dickinson, J. C., Jr. 1953. Report on the McCabe collection of British Columbian birds. Bull. Mus. Comp. Zool., 109:(2).
Dixon, Keith L. 1955. An ecological analysis of the interbreeding of crested titmice in Texas. Univ. Calif. Publ. Zool., 54:125-206.
Dorst, J. 1947. Revision systematique du genre Corvus. L'Oiseau et la Rev. Franc. D'Ornith., 17:44-87.
Dwight, Jonathan, Jr. 1893. Summer birds of Prince Edward Island. Auk, 10:1-15.
Eaton, E. H. 1923. Birds of New York. N. Y. State Museum, Albany. 719 pp.

Literature Cited

Emlen, J. T., Jr. 1936. Age determination in the American Crow. Condor, 38:99-102.

──────. 1940. The midwinter distribution of the crow in California. Condor, 42:287-94.

──────. 1942. Notes on a nesting colony of western crows. Bird-Banding, 13:143-54.

Fay, F. H., and T. J. Cade. 1959. An ecological analysis of the avifauna of St. Lawrence Island, Alaska. Univ. Calif. Publ. Zool., 63:73-150.

Fitch, Henry S. 1958. Home ranges, territories, and seasonal movements of vertebrates of the Natural History Reservation. Univ. Kans. Publ., Mus. Nat. Hist., 11:63-326.

Forbush, E. H. 1925-29. Birds of Massachusetts and other New England States. 3 vols. Mass. Dept. of Agric., Boston.

Frings, Hubert, and Mable Frings. 1959. The language of crows. Scientific American, 201:119-31.

Frings, Hubert, Mable Frings, Joseph Jumber, Jacques Giban, René-Guy Busnes, Phillippe Gramet. 1958. Reactions of American and French species of Corvus and Larus to recorded communication signals tested reciprocally. Ecology, 39:126-31.

Gabrielson, I. N., and S. G. Jewett. 1940. Birds of Oregon. Ore. State Monog., Studies in Zool., No. 2, 650 pp.

Gabrielson, I. N., and Frederick C. Lincoln. 1959. The Birds of Alaska. The Stackpole Co., Harrisburg, Pa. 922 pp.

Goldman, Edward A. 1951. Biological investigations in Mexico. Smiths. Misc. Coll., 115:1-476.

Good, Ernest E. 1952. "The life history of the American Crow Corvus brachyrhynchos Brehm." Unpublished Ph. D. thesis, Ohio State University, Columbus, Ohio. 190 pp.

Gosse, Philip H. 1847. The Birds of Jamaica. London.

Greene, Earle R. 1946. Birds of the Lower Florida Keys. Quart. Jour. Fla. Acad. Sci., 8.

Griffin, William W. 1940. Nesting extremes for birds breeding in the Atlanta region. Oriole, 5:1-6.

Grinnell, Joseph, and Alden H. Miller. 1944. The distribution of the birds of California. Pac. Coast Avif., 27:1-608.

Griscom, Ludlow, and Dorothy E. Snyder. 1955. The Birds of Massachusetts. Peabody Museum, Salem, Mass. 295 pp.

Harper, Francis. 1953. Birds of the Nueltin Lake Expedition, Keewatin, 1947. Amer. Midl. Nat., 49:1-116.

──────. 1958. Birds of the Ungava Peninsula. Univ. Kans. Misc. Publ., 17:1-171.

Hebard, Frederick V. 1941. Winter birds of the Okefinokee and Coleraine. Ga. Soc. Nat., Bull. No. 3. 84 pp.
Hellmayr, Charles E. 1934. Catalogue of birds of the Americas and the adjacent islands. Field Mus. Nat. Hist., 13:7.
Howell, A. H. 1913. Descriptions of two new birds from Alabama. Proc. Biol. Soc. Wash., 26:199-200.
—————. 1928. Birds of Alabama. Dept. of Game and Fisheries of Alabama. 384 pp.
—————. 1932. Florida Bird Life. Coward-McCann, Inc., New York. 479 pp.
Jewett, S. G., and I. N. Gabrielson. 1929. Birds of the Portland area, Oregon. Pac. Coast Avif., 19:54 pp.
Jewett, S. G., W. P. Taylor, W. T. Shaw, and J. W. Aldrich. 1953. Birds of Washington State. University of Washington Press, Seattle, Wash. 767 pp.
Johnson, David H., Monroe D. Bryant, and Alden H. Miller. 1948. Vertebrate animals of the Providence Mountains area of California. Univ. Calif. Publ. Zool., 48:221-376.
Johnston, David W. 1947. Fish Crow and Tennessee Warbler at Athens. Oriole, 12:33-34.
—————. 1959. The incubation patch and related breeding data of crows. Murrelet, 40:6.
Johnston, David W., and T. P. Haines. 1957. Analysis of mass bird mortality in October, 1954. Auk, 74:447-58.
Kellogg, Peter Paul and Robert C. Stein. 1953. Audio-spectrographic analysis of the songs of the Alder Flycatcher. Wils. Bull., 65:75-80.
Kitchin, E. A. 1949. Birds of the Olympic Peninsula. Olympic Stationers, Port Angeles, Wash. 262 pp.
Laing, H. M. 1925. Birds collected and observed during the cruise of the *Thiepval* in the north Pacific, 1924. Victoria Mem. Mus. Bull., 40:1-46.
—————. 1942. Birds of the coast of Central British Columbia. Condor, 44:177.
Lanyon, Wesley E. 1957. The comparative biology of the meadowlarks *(Sturnella)* in Wisconsin. Publ. Nuttall Ornith. Club, 1:1-67.
Lanyon, Wesley E., and William R. Fish. 1958. Geographical variation in the vocalizations of the Western Meadowlark. Condor, 60:339-41.
MacArthur, Robert H., and Robert A. Norris. A method of pooling standard-deviation values in combinable samples. In press.

Literature Cited

Marler, Peter, and Donald Isaac. 1960. Physical analysis of a simple bird song as exemplified by the Chipping Sparrow. Condor, 62:124-35.

Marshall, Joe T., Jr. 1948. Ecologic races of song sparrows in the San Francisco Bay region. Part I, Condor, 50:193-215. Part II, Condor, 50:233-56.

Mathews, F. Schuyler. 1921. Field Book of Wild Birds and Their Music. G. P. Putnam's Sons, New York. 401 pp.

Mayr, Ernst. 1942. Systematics and the Origin of Species. Columbia University Press, New York. 334 pp.

—————. 1946. History of the North American bird fauna. Wils. Bull., 58:3-41.

—————. 1956. Gesang und Systematik. Beitrage zur Vogelkunde, V:112-17.

Mayr, E., and D. Amadon. 1951. A classification of recent birds. Amer. Mus. Nov., 1496:1-42.

Mayr, E., E. G. Linsley, and R. L. Usinger. 1953. Methods and Principles of Systematic Zoology. McGraw-Hill Book Co., Inc., New York. 336 pp.

Meinertzhagen, R. 1926. Introduction to a review of the genus *Corvus*. Nov. Zool., 33:57-121.

Meise, W. 1928. Die Verbreitung der Aaskrähe (Formenkreis *Corvus corone* L.). J. Ornith., 76:1-203.

Miller, Alden H. 1951. An analysis of the distribution of the birds of California. Univ. Calif. Publ. Zool., 50:531-644.

Miller, Alden H., ed. 1957. Distributional Check-List of the Birds of Mexico. Pac. Coast Avif. 33, Part II. 435 pp.

Miller, Robert C., and Elizabeth Curtis. 1940. Birds of the University of Washington campus. Murrelet, 21:34-46.

Munro, J. A., and I. McT. Cowan. 1947. A review of the Bird Fauna of British Columbia. Spec. Pub. No. 2, British Columbia Provincial Museum, Victoria. 285 pp.

Murray, J. J. 1952. A check-list of the birds of Virginia. Va. Soc. Ornith:74-75.

Nice, M. M., and L. B. Nice. 1924. The birds of Oklahoma. Univ. Okla. Bull., 20:1-122.

Norris, Robert A. 1958. Comparative biosystematics and life history of the nuthatches *Sitta pygmaea* and *Sitta pusilla*. Univ. Calif. Publ. Zool., 56:119-300.

Palmer, Ralph S. 1949. Maine birds. Bull. Mus. Comp. Zool., 102:1-656.

Pearse, Theed. 1946. Comox, Vancouver Island—1917-1944. Murrelet, 27:7.

Pearson, T. Gilbert. 1922. Notes on the birds of Cumberland Island, Georgia. Wils. Bull., 34:84-90.
Pearson, T. G., C. S. Brimley, and H. H. Brimley. 1942. Birds of North Carolina. N. C. Dept. of Agric., Raleigh. 380 pp.
Peters, Harold S. 1936. A list of external parasites from birds of the eastern part of the United States. Bird-Banding, 7:9-27.
Peters, H. S., and Thomas D. Burleigh. 1951. The Birds of Newfoundland. Dept. Nat. Resources, St. John's, Province of Newfoundland. 450 pp.
Peterson, Roger T. 1958. A Field Guide to the Birds. Houghton Mifflin Co., Boston. 290 pp.
Phillips, Allan R. 1942. A new crow from Arizona. Auk, 59: 573-75.
Pitelka, Frank A. 1951. Speciation and ecologic distribution in American jays of the genus Aphelocoma. Univ. Calif. Publ. Zool., 50:195-464.
Preble, Edward A. 1908. A biological investigation of the Athabaska-MacKenzie region. N. Am. Fauna, 27. 574 pp.
Ressel, C. B. 1889. Birds of Chester County, Penn. Ornith. and Ool., 14:101.
Ridgway, Robert. 1904. The Birds of North and Middle America. U. S. Nat. Mus. Bull., 50, Part III.
Sage, John H., Louis B. Bishop, and Walter P. Bliss. 1913. The Birds of Connecticut. State of Conn. State Geol. and Nat. Hist. Surv. Bull. No. 20. P. 108.
Schorger, A. W. 1941. The crow and the raven in early Wisconsin. Wils. Bull., 53:103-6.
Selander, Robert K., and Donald R. Giller. 1959. Interspecific relations of woodpeckers in Texas. Wils. Bull., 71:107-24.
Sibley, Charles G. 1957. The evolutionary and taxonomic significance of sexual dimorphism and hybridization in birds. Condor, 59:166-91.
Sprunt, Alexander, Jr. 1954. Florida Bird Life. Coward-McCann, Inc., New York. 527 pp.
Sprunt, Alexander, Jr., and E. Burnham Chamberlain. 1949. South Carolina Bird Life. University of South Carolina Press, Columbia, S. C. 585 pp.
Stewart, Robert E., and Chandler S. Robbins. 1958. Birds of Maryland and the District of Columbia. N. Am. Fauna, 62. 401 pp.
Sumner, Lowell and Joseph S. Dixon. 1953. Birds and Mammals of the Sierra Nevada. University of California Press, Berkeley. 484 pp.

Literature Cited

Sutton, George M. 1951. Mexican Birds. First Impressions. University of Oklahoma Press, Norman, Okla. 297 pp.
Sutton, George M., and O. S. Pettingill, Jr. 1942. Birds of the Gomez Farias Region, southwestern Tamaulipas. Auk, 59:23.
Taverner, P. A. 1919. Birds of Eastern Canada. Can. Dept. Mines, Memoir 104, No. 3. Biol. Ser. 297 pp.
─────. 1926. Birds of Western Canada. Victoria Mem. Mus. Bull. No. 41, 259-62.
Taylor, L. 1955. Introduction to the Birds of Jamaica. Macmillan and Co., Ltd., London. 114 pp.
Teal, John M. 1959. Birds of Sapelo Island and vicinity. Oriole, 24:1-14.
van Rossem, A. J. 1945. A distributional survey of the birds of Sonora, Mexico. Occas. Papers, Mus. Zool., Louisiana State Univ., 21:1-379.
Van Tyne, J., and A. J. Berger. 1959. Fundamentals of Ornithology. John Wiley and Sons, Inc., New York. 624 pp.
Vaurie, Charles. 1957. Field notes on some Cuban birds. Wils. Bull., 69:301-13.
─────. 1958. Remarks on some *Corvidae* of Indo-Malaya and the Australian Region. Amer. Mus. Nov., 1915:1-13.
Wayne, Arthur T. 1910. Birds of South Carolina. Contr. Charleston Mus. I. 254 pp.
Wetmore, Alexander. 1927. The birds of Porto Rico and the Virgin Islands. New York Acad. Sci., 9:409-598.
─────. 1956. A check-list of the fossil and prehistoric birds of North America and the West Indies. Smiths. Misc. Coll., 131:1-105.
Wetmore, Alexander, and Bradshaw H. Swales. 1931. The birds of Haiti and the Dominican Republic. U. S. Nat. Mus. Bull. 155:1-483.
Wilhelm, Eugene J., Jr. Extension in breeding range of the Fish Crow. In press.
Wolfe, L. R. 1956. Check-list of the Birds of Texas. Intelligencer Printing Co., 89 pp.
Youngworth, William. 1935. The birds of Fort Sisseton, South Dakota, a sixty year comparison. Wils. Bull., 47:209-35.
Zimmerman, Dale A. 1957. Notes on Tamaulipan birds. Wils. Bull., 69:273-77.

INDEX

Age characteristics: of Common Crow, 8, 18-19; of Australian crows, 82; of Cuban Crow, 82; of Jamaican Crow, 98
Asian land bridge, 103, 104
Audiospectrograms: Common Crow, 15, 63; Fish Crow, 63; Mexican Crow, 73

Banded crows, 10
Behavior: as a method of identifying, 7; in subspecies of Common Crow, 36
Biosystematics, 4

Chattering Crow. See *Corvus jamaicensis*
Color: as means of identifying, 6, 7. See also individual species' accounts
Common Crow. See *Corvus brachyrhynchos*
Crows *americanus*. See *Corvus brachyrhynchos*
Corvus brachyrhynchos: geographic range of, 12, 13, 20, 21; habitat of, 13, 14, 62, 65; breeding of, 14, 65, 66; voice of, 15, 16, 63, 91; morphology of, 16; subspecies of, 17, 18, 76; food of, 62, 63
Corvus brachyrhynchos brachyrhynchos: distribution of, 17, 25, 26; winter specimens, 24; measurements of, 25; comparison with other subspecies, 25, 26
Corvus brachyrhynchos caurinus: range of, 13, 27, 36; nesting habits of, 27, 35; measurements of, 28, 33, 34, 35; voice of, 29, 30, 31, 32; habitat choice of, 32, 37; fossils of, 35; behavior of, 36; type specimens of, 37
Corvus brachyrhynchos hargravei: distribution of, 18, 21; comparison with other populations, 25; measurements of, 25
Corvus brachyrhynchos hesperis: distribution of, 17, 26; measurements and size of, 18, 25, 26; comparison with *brachyrhynchos*, 25, 26; comparison with *caurinus*, 28, 30, 31, 32, 34, 35, 36; voice of, 30, 31
Corvus brachyrhynchos pascuus: distribution of, 17; comparison with other subspecies, 22, 23
Corvus brachyrhynchos paulus: distribution of, 17, 19, 21; comparison with other subspecies, 18, 23, 24, 25
Corvus caurinus. See *Corvus brachyrhynchos caurinus*
Corvus corone: voice of, 5, 91; systematics of, 5, 7, 22, 103
Corvus cryptoleucus, 82, 87
Corvus imparatus: comparison with other forms, 64, 74, 75; geographic range of, 71; habitat of, 71, 75; voice of, 72, 73, 74; seasonal color changes of, 74; morphology of, 74, 75, 76, 77, 98; eastern and western populations, 75, 76, 77; subspecies of, 77
Corvus imparatus imparatus, 77, 106

Corvus imparatus sinaloae, 77, 106
Corvus jamaicensis: geographic range of, 97; habitat of, 97; voice of, 97; morphology of, 97, 98
Corvus leucognaphalus: geographic range of, 85; habitat of, 85, 86, 87; subspecies, 86; comparison with other forms, 86, 87; morphology of, 86, 87; voice of, 86, 91
Corvus leucognaphalus erythrophthalmus, 86
Corvus minutus, 90, 91, 93
Corvus nasicus: geographic range of, 81; habitat of, 81; voice of, 81, 82, 97; comparison with other forms, 82; morphology of, 82; mallophaga on, 102
Corvus ossifragus: voice of, 30, 63, 64, 91; geographic range of, 60; range extensions of, 60, 61; habitat choice of, 60, 61, 62, 65, 75, 92; breeding of, 61, 62, 65, 66; food of, 62, 63; colors of, 63; comparison with other species, 64, 92
Corvus ossifragus caurinus, 29
Corvus ossifragus palmarum, 91, 92
Corvus palmarum: comparison with other forms, 64, 91, 92; geographic range of, 90; habitat of, 90, 91, 92; morphology of, 91, 92, 93, 94; subspecies, 92, 93, 94; seasonal color differences of, 93, 94; mallophaga on, 102
Corvus pumilis, 85, 104
Cuban Crow. See *Corvus nasicus*

Eastern Crow. See *Corvus brachyrhynchos brachyrhynchos*
Ectoparasites, 8, 102
European Carrion Crow. See *Corvus corone*
European Rook, 97

Fish Crow. See *Corvus ossifragus*
Florida Crow. See *Corvus brachyrhynchos pascuus*
Fossil crows, 35

Habitat choice: as a method of identifying, 7; of White-necked Raven, 87; of North American crows, 104. See *also* individual species' accounts

Haitian Crow, 86
Host-parasite relationships, 102
Hybrids: between *caurinus* and *hesperis,* 28, 34, 35; absence of, 64, 67

Isolating mechanisms: general, 4, 73; between Fish and Common Crows, 64-66; of Mexican Crow, 76-77; of North American crows, 104

Jabbering Crow. See *Corvus jamaicensis*
Jamaican Crow. See *Corvus jamaicensis*

Mallophaga, 102
Measurements: of first-year birds, 8, 9; as a method of identifying, 68. See *also* individual species' accounts
Mexican Crow. See *Corvus imparatus*
Migration: as a method of identifying, 8; of Mexican and Caribbean Crows, 10; of Common Crows, 26

North American crows: size and color groups of, 102, 103, 104, 107; spread of, 104
Northwestern Crow. See *Corvus brachyrhynchos caurinus*

Palm Crow. See *Corvus palmarum*
Phylogeny, 102-4

Ravens, 86, 87, 103
Reproductive isolation: in *caurinus,* 36; between Fish and Common Crows, 65-67
Reproductive phenology: as a method of identifying, 7; between Fish and Common Crows, 66-67

Scale-like effect: as a method of identifying, 7; in Common Crow, 16-17; in Fish Crow, 17, 64; in Mexican Crow, 74; in Cuban Crow, 82, 87; in White-necked Crow, 87; in Jamaican Crow, 98
Seasonal isolation, 65-67
Selection of specimens, 8-10, 18-19
Seventy-five per cent rule, 19

Song sparrow, ecological isolation of, 36-37
Southern Crow. See *Corvus brachyrhynchos paulus*
Statistics: formulae, 9; interpretation of, 19-20
Sympatry: between *caurinus* and *hesperis,* 27-28, 34; between *ossifragus* and *brachyrhynchos,* 62, 64; between *palmarum* and *leucognaphalus,* 91; between *palmarum* and *nasicus,* 91; in North American crows, 104

Voice: as a method of identifying, 7. See also individual species' accounts

Weights: as a method of identifying, 7; of Common Crow, 100; use in taxonomy, 100
Western Crow. See *Corvus brachyrhynchos hesperis*
White-necked Crow. See *Corvus leucognaphalus*
White-necked Raven. See *Corvus cryptoleucus*
Winter specimens, 10, 24
Zone of intergradation, 31-37